Encounter
Physical Geography

INTERACTIVE EXPLORATIONS OF EARTH
Using Google Earth™

JESS C. PORTER
University of Arkansas at Little Rock

STEPHEN O'CONNELL
University of Central Arkansas

PEARSON

Boston   Columbus   Indianapolis   New York   San Francisco   Upper Saddle River
Amsterdam   Cape Town   Dubai   London   Madrid   Milan   Munich   Paris   Montréal   Toronto
Delhi   Mexico City   São Paulo   Sydney   Hong Kong   Seoul   Singapore   Taipei   Tokyo

Geography Editor: Christian Botting

Senior Marketing Manager: Maureen McLaughlin

Assistant Editor: Kristen Sanchez

Media Producer: Ziki Dekel

Editorial Assistant: Bethany Sexton

Senior Marketing Assistant: Nicola Houston

Managing Editor, Geosciences and Chemistry: Gina M. Cheselka

Production Project Manager: Ed Thomas

Cover Design: Seventeenth Street Studios

Operations Specialist: Michael Penne

Front cover image credit: NASA.

Back cover left: Data SIO, NOAA, U.S. Navy, NGA GEBCO / © 2013 Cnes/Spot Image / Image © 2013 TerraMetrics.

Back cover right: Image © 2013 TerraMetrics /© 2013 Cnes/Spot Image.

Access card top: IBACAO / Data SIO, NOAA, U.S. Navy, NGA, SEBCO / 2013 Snes/Spot Image / Image © 2013. TerraMetrics

Access card bottom: © 2012 Terra Metrics / Image USDA Farm Service Agency / Image © 2012 GeoEye.

Credits and acknowledgments for materials borrowed from other sources and reproduced, with permission, in this textbook appear on the appropriate page within the text.

www.pearsonhighered.com

1 2 3 4 5 6 7 8 9 10—EBM—17 16 15 14 13

ISBN-10: 0-321-67252-6; ISBN-13: 978-0-321-67252-0

Contents

Preface   iv

Chapter 1: Introduction to Google Earth™   1

Chapter 2: Portraying Earth   15

Chapter 3: Introduction to the Atmosphere   31

Chapter 4: Insolation and Temperature   45

Chapter 5: Atmospheric Pressure and Wind   61

Chapter 6: Atmospheric Moisture   75

Chapter 7: Atmospheric Disturbances   89

Chapter 8: Climate and Climate Change   103

Chapter 9: The Hydrosphere   117

Chapter 10: Cycles and Patterns in the Bisosphere   133

Chapter 11: Terrestrial Flora and Fauna   147

Chapter 12: Soils   161

Chapter 13: Introduction to Landform Study   175

Chapter 14: The Internal Processes   189

Chapter 15: Weathering and Mass Wasting   203

Chapter 16: Fluvial Processes   217

Chapter 17: Karst and Hydrothermal Processes   231

Chapter 18: The Topography of Arid Lands   245

Chapter 19: Glacial Modification of Terrain   259

Chapter 20: Coastal Processes and Terrain   273

Preface

Welcome to *Encounter Physical Geography*! This workbook will immerse you in interactive explorations of the world's immense diversity of physical processes and landscapes. Elements of geography and geospatial techniques come together to give you a better understanding of our world's variations and commonalities. This is accomplished by applying the power of the Google Earth™ program to zoom in and around features and landscapes ranging from street-corner scenes to regional-scale patterns. Within Google Earth™, we will utilize associated tools and layers such as photographs, satellite imagery, and historic maps. We will springboard into related websites that help us understand the patterns and processes at work on our dynamic planet.

As you work through the explorations contained in this book, you will feel the "big picture" understanding of the world come into focus. Not only will your knowledge of the themes of geography grow, but also your ability to apply these themes to your interpretation and understanding of life on Earth. While these exercises are designed to educate thematically and topically, they are also designed to be fun. We encourage you to take these exercises beyond the parameters laid out in the multiple-choice and short-answer question segments of the workbook. If something piques your curiosity, dig-in deeper. Look for the answers, but also look for more questions to ask. It is our hope that you will apply your enhanced spatial understanding of the world and its physical attributes beyond your studies associated with this workbook. You will find that improving your spatial thinking skills is something that can benefit every aspect of your life.

Google Earth™ can be downloaded for free from http://earth.google.com. The Google Earth™ website has a wide variety of instructional material from tutorials to demonstration videos. You can download interesting collections of spatial data in KMZ files. Explore the software and take advantage of Google's ancillary material to familiarize yourself with your new aid to geographic learning.

The initial two chapters of this workbook address core geographic concepts while introducing you to the functionality of Google Earth™. Location, scale, and place are discussed and the basics of interpreting remotely sensed images are introduced. This will help you better understand and interpret what you are seeing in the Google Earth™ environment. The workbook then examines 18 significant elements of physical geography beginning with atmospheric composition.

Each of these chapters is organized into four subtopics where you will have the opportunity to answer multiple-choice and short-answer questions. These questions have been designed with an emphasis on high-level assessment skills that will encourage you to interpret and appraise the imagery and information uncovered. You will utilize Pearson's companion website, www.mygeoscienceplace.com, to download KMZ files, worksheets, and electronic versions of the assessments.

We are thankful for the hard work of editors and staff at Pearson Geography in the development and production of Encounter Physical Geography. Specific thanks must go to Christian Botting and Kristen Sanchez, who have continued to provide exceptional project leadership and support. We are also appreciative of the accuracy reviewers for the project, Miriam Helen Hill and Maura Abrahamson. Finally, we wish to express our gratitude to our respective families. Over the duration of this project they have exhibited patience and offered encouragement.

Jess C. Porter
jcporter@ualr.edu

Stephen M. O'Connell
soconnell@uca.edu

Encounter Physical Geography

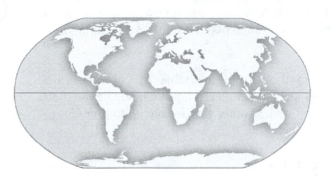

Name:_____

Date: _____

Chapter 1

Introduction to GoogleEarth™

Exploration 1.1: Getting Started with Navigation
Multiple Choice

This chapter will introduce you to Google Earth™. Google Earth™ is a computer application that provides opportunities to see the Earth's varied geography from a range of aerial and street-level perspectives. You can zoom in and out, rotate, and tilt your perspective of the Earth. You will learn how to navigate to different locations and use some of the tools that this powerful application provides. Note: Google Earth™ is an application that is updated frequently. As a result, the functions and appearance of the application may vary slightly from what is described.

Instructions for all Parts:

1. To begin, verify that your computer has the latest version of Google Earth™ installed. If you do not, then go to http://earth.google.com and download the most recent version of the free Google Earth™ software. Your computer will need a high-speed Internet connection in order to utilize the application effectively.

2. Make sure you have opened the **KMZ** file from www.mygeoscienceplace.com and open it in GoogleEarth.

3. From the Places panel, expand **1. Introduction to Google Earth.kmz** then open the **1. Introduction to Google Earth** folder. The **1.1 Getting Started with Navigation** folder includes two web links you will use to complete this exploration.

4. The very best way to learn Google Earth™ is to dive in and explore your world. The software is intuitive to use and not complicated. However, you will also find it useful to learn some of the basics of navigation as demonstrated in the Google Earth™ User Guide. Go to http://www.google.com/earth/learn/ for some quick tutorials regarding the use of this software. Do not forget about these useful resources. For more specific information you can utilize the Help Resources found at http://support.google.com/earth or by clicking Help, then Help Resources from the menu bar at the top of the Google Earth™ window. If you are ever confused or need help with using the software, these sites will be your best bet for quick and easy assistance.

5. In the Help Resources, under the Getting started and basics heading is a link titled Basic features user guide. Open this link and study the Getting to Know Google Earth™

diagram and associated information. Be sure to click the links associated with the tools to explore the capabilities of Google Earth™. Without spending some time exploring the Help Resources and tutorials, you may be unable to answer the following questions and will likely encounter significant difficulties in your attempts to complete the explorations of Encounter Physical Geography.

6. Now click the Navigation link, located to the left in the table of contents. Click "Navigating in Google Earth" and play the short video to learn how to navigate the globe in Google Earth™. Scroll down the page and review other tips for navigating in Google Earth™.

Exploration 1.1 Getting Started with Navigation Part A

A. If you needed to find a specific place on Earth, you would use the:

1. Places panel
2. Layers panel
3. Overview map
4. Finder tool
5. Search panel

Exploration 1.1 Getting Started with Navigation Part B

B. On what part of the Google Earth™ window are the on-screen navigation controls located?

1. bottom right
2. top right
3. center
4. bottom left
5. top left

Exploration 1.1 Getting Started with Navigation Part C

C. Which of the following is **not** a capability of Google Earth™?

1. You can display sunlight across the landscape.
2. You can add polygons, lines, and points (placemarks) to the view.
3. You can view live imagery.
4. You can display historical imagery.
5. You can email views from Google Earth™

Exploration 1.1 Getting Started with Navigation Part D

D. Where would you find the information regarding a view's coordinates, elevation, and imagery date?

1. In the Places panel
2. In the Search panel
3. In the Layers panel
4. At the bottom of the 3D Viewer
5. In the Overview map

Exploration 1.1 Getting Started with Navigation
Short Answer

Instructions for all Parts:

1. To begin, verify that your computer has the latest version of Google Earth™ installed. If you do not, then go to http://earth.google.com and download the most recent version of the free Google Earth™ software. Your computer will need a high-speed Internet connection in order to utilize the application effectively.

2. Make sure you have opened the **KMZ** file from www.mygeoscienceplace.com

3. From the Places panel, expand **1. Introduction to Google Earth.kmz** then open the **1. Introduction to Google Earth** folder. The **1.1 Getting Started with Navigation** folder includes two web links you will use to complete this exploration.

4. The very best way to learn Google Earth™ is to dive in and explore your world. The software is intuitive to use and not complicated. However, you will also find it useful to learn some of the basics of navigation as demonstrated in the Google Earth™ User Guide. Go to http://www.google.com/earth/learn/ for some quick tutorials regarding the use of this software. Do not forget about these useful resources. For more specific information you can utilize the Help Resources found at http://support.google.com/earth or by clicking Help, then Help Resources from the menu bar at the top of the Google Earth™ window. If you are ever confused or need help with using the software, these sites will be your best bet for quick and easy assistance.

5. In the Help Resources, under the Getting started and basics heading is a link titled Basic features user guide. Open this link and study the Getting to Know Google Earth™ diagram and associated information. Be sure to click the links associated with the tools to explore the capabilities of Google Earth™. Without spending some time exploring the Help Resources and tutorials, you may be unable to answer the following questions and will likely encounter significant difficulties in your attempts to complete the explorations of Encounter Physical Geography.

6. Now click the Navigation link, located to the left in the table of contents. Click "Navigating in Google Earth" and play the short video to learn how to navigate the globe in Google Earth™. Scroll down the page and review other tips for navigating in Google Earth™.

Instructions for Exploration 1.1 Getting Started with Navigation Short Answer A:

1. Enter the location of your university in the Search panel and then view the image of your university.

Exploration 1.1 Getting Started with Navigation Short Answer A

A. Seeing your university from an aerial perspective, describe some insight that you have gained. Did you find buildings you did not know existed? Maybe a parking lot is closer to your dorm than the one you currently use? If you are an online student, how is the campus different than what you envisioned? Feel free to be creative in your response.

_____

_____

_____

_____

_____

_____

_____

Instructions for Exploration 1.1 Getting Started with Navigation Short Answer B

1. One can navigate in Google Earth™ using either on-screen controls, the mouse, the keyboard, or any combination of these approaches. To see what keyboard commands perform what tasks, click the "Navigating with the keyboard" link in the "Getting started and basics" section of the User Guide. Note that the keystrokes differ for a Mac and a PC.

Exploration 1.1 Getting Started with Navigation Short Answer B

B. Navigate using both the on-screen controls such as the zoom slide and move joystick; then navigate using the keyboard and/or mouse. Try integrating the two approaches. Which method(s) do you prefer and why?

_____

_____

_____

_____

_____

_____

_____

Exploration 1.2 – Layers
Multiple Choice

Google Earth's™ functionality is magnified by the ability to add layers of data and information to the 3D Viewer. Beyond the visual representation of the Earth's surface we can add any number of features such as the road networks, place names, boundaries, photographs, business locations, current weather conditions, and three-dimensional buildings to name just a few. A problem that faces many users of Google Earth™, however, is that their viewers become cluttered with too much information. In general, it is best practice to simply display only the layers you need at any given time. *Borders and Labels* is the layer that you will find most useful without cluttering the view.

Instructions for all Parts:

1. Make sure you have opened the **KMZ** file from www.mygeoscienceplace.com

2. From the Places panel, expand **1. Introduction to Google Earth.kmz** then open the **1. Introduction to Google Earth** folder. Open the **1.2 Layers** folder.

3. Double-click the *Albuquerque, NM*, placemark. Now check the boxes beside the ten or so layers and/or folders in the *Primary Database* in the Layers pane beneath the Places pane. Wait a few moments and you will get a taste of how cluttered Google Earth™ can become. It will take several moments for everything to load.

4. Now turn off all of the layers in the *Primary Database* except *Borders and Labels*. Depending on the type of question you seek to answer, you may want to activate one or several folders or layers at a time.

5. For example, let's start with the *Photos* layer. Turn this layer of information on by checking the box next to *Photos* in the *Primary Database*. You will see a number of small blue and red squares appear on the screen.

6. Begin clicking the blue squares and you will see that these are photographs of features that are correlated with their location on the ground. Click on approximately ten of these boxes contained in the current view.

7. Now click one of the red squares to see 360-degree views that are available. When these images are activated, you can click the photograph to fly into the image.

8. Click Exit Photo in the upper-right corner to return to the 3D Viewer. Return to the original perspective by double-clicking the *Albuquerque, NM*, placemark.

Exploration 1.2 Layers Part A

A. What sphere is indicated by the curvilinear feature at the *Albuquerque, NM*, placemark?

1. Lithosphere
2. Hydrosphere
3. Biosphere
4. Atmosphere
5. Cryosphere

Instructions for Exploration 1.2 Layers Part B:

1. Turn off all of the layers in the *Primary Database* except the *Roads* layer. You will see that the major highways of the community are illuminated and labeled. This *Roads* layer can be very helpful when you are attempting to analyze human patterns on the landscape because transportation is closely related to many types of development.

2. Now use the zoom-slider in the Google Earth™ navigation panel or the scroll wheel on your mouse to slowly zoom in. As you zoom in, you will notice that more and more details of the road network will emerge until the smallest residential streets appear and are labeled. Zoom out and these features and labels disappear.

Exploration 1.2 Layers Part B

B. Identify the road that crosses the bridge immediately to the north of the *Albuquerque, NM*, placemark.

1. Coors Boulevard
2. Interstate 40 (Coronado Freeway)
3. Central Avenue
4. State Highway 314
5. Montano Road

Instructions for Exploration 1.2 Layers Part C:

1. Geographic information systems (GIS) are all about added information like the *Roads* layer. We can take layers such as a road network and do analysis in relation to the built environment. Throughout this text you will be capitalizing on the ability of Google Earth™ to display and overlay different types of data and information on the surface of the digital globe in order to increase your understanding of the world, physical, and human geography.

2. Turn off all layers in the *Primary Database* except the *3D Buildings* layer.

3. Double-click the *Downtown* placemark.

Exploration 1.2 Layers Part C

C. What is the name of the building in downtown Albuquerque that would cast the longest shadow?

1. Albuquerque/Bernalillo Government Center
2. Albuquerque Convention Center
3. Albuquerque Court House
4. Albuquerque Plaza / Hyatt Regency
5. Albuquerque Petroleum Building

Instructions for Exploration 1.2 Layers Part D:

1. Turn off all layers except *Roads* in the *Primary Database*. Street-view imagery is increasingly available in urban locations around the world. This tool can really do a great job of providing you with a sense of place of a particular location as you can simulate driving down a road.

2. Double-click the *Downtown* placemark. Locate Central Avenue near the yellow push-pin for downtown. Continue to zoom in toward the street near the placemark until the screen switches to Street-view.

3. Notice that can advance down the street by clicking on the yellow line or using your mouse's scroll wheel. Exit street view and zoom out to a broad view of Albuquerque.

4. Double-click the *Southern Boulevard* placemark and examine the street view imagery at the location.

5. Double-click the *Sandia Crest Road* placemark and examine the street view imagery at the location

Exploration 1.2 Layers Part D

D. Utilizing the street view imagery from the two greater Albuquerque area placemarks (Southern Boulevard and Sandia Crest Road), which of the following statements is most strongly supported?

1. There is significant variation in the vegetation of the greater Albuquerque area.
2. The greater Albuquerque area has uniform vegetation.
3. The greater Albuquerque area does not have street view imagery available.
4. The greater Albuquerque area is entirely flat.
5. It is never cloudy in greater Albuquerque.

When you complete this exploration, turn off the **1.2 Layers** folder.

Exploration 1.2 Layers
Short Answer

Instructions for Exploration 1.2 Layers Short Answer A:

1. Turn off all layers except *Photos* in the *Primary Database*. Internet users can upload images to specific locations and plot these in Google Earth™. Utilizing these images can help you get a better understanding of the local landscape of a given location.

2. Use the Search panel to fly to Albuquerque, New Mexico.

Exploration 1.2 Layers Short Answer A

A. Based solely on what you can ascertain from your sampling of images in the *Photos* layer, write a short paragraph that describes the physical landscape and community of Albuquerque, New Mexico. Be sure to mention vegetation characteristics and the likely type of climate.

_____

_____

_____

_____

_____

_____

Instructions for Exploration 1.2 Layers Short Answer B

1. Turn off all layers except *Borders and Labels* layer and the *Clouds* and *Radar* layers from the *Weather* folder in the *Primary Database*.

2. Change your view so you can see the full extent of North America.

Exploration 1.2 Layers Short Answer B

B. Describe the clouds and precipitation patterns over North America. Is the correlation between cloud cover and precipitation strong? Can you identify a storm system (area of low pressure)? Where are these located? Has or will this storm system affect you at your current location? Document the date and time you viewed these features in your response.

_____

_____

_____

_____

_____

_____

_____

Exploration 1.3 Location
Multiple Choice

Google Earth™ is a terrific tool to visualize Earth's complex physical landscapes. The three-dimensional capacity of the program allows the user to get a grasp on the variations in relief that a paper-map is not able to provide.

Instructions for all Parts:

1. Make sure you have opened the **KMZ** file from the blue box on the left.

2. From the Places panel, expand **1. Introduction to Google Earth.kmz** then open the **1. Introduction to Google Earth** folder. Open and turn on the **1.3 Location** folder.

3. Turn off all layers in the *Primary Database*.

4. Open the Google Earth™ Options dialog box by selecting Tools > Options from the Menu bar. On the 3D View tab, verify that elevation is displayed using Meters, Kilometers.

Instructions for Exploration 1.3 Location Part A:

1. Double-click the *Grand Canyon* placemark. You will zoom to a view of this unique natural feature. As you move the cursor across the screen you will see the values for elevation that are displayed at the bottom of the screen change.

Exploration 1.3 Location Part A

A. What is the approximate elevation value of the Colorado River as displayed in the *Grand Canyon* placemark?

1. 475 meters above sea level
2. 1145 meters above sea level
3. 1550 meters above sea level
4. 2400 meters above sea level
5. 6335 meters above sea level

Instructions for Exploration 1.3 Location Parts B–D:

1. Double-click the *Point of Origin* placemark. What is significant about this location off the west coast of Africa? This is the origin for our globe's predominant location system. This is where the Earth's north/south baseline (the equator) and the Earth's east/west baseline (the Prime Meridian/International Date Line) intersect. Let's make it a little easier to see.

2. Click View then click Grid. The latitude and longitude grid is illuminated. You can see that the *Point of Origin* placemark is located at 0° latitude and 0° longitude.

3. Click Tools and then Options. In the Show Lat/Long box, you will see four options for displaying your absolute location. Select Decimal Degrees and then click OK.

4. At the center bottom of your screen you will see the coordinates for the location of your cursor (open hand icon). Examine what happens to the latitude and longitude numbers as you move the cursor between the southern and northern hemispheres and the eastern and western hemispheres. A positive latitude number represents locations north of the equator while a positive longitude represents locations east of the Prime Meridian. Move your cursor across the grid again to verify these statements. Now manipulate the globe by rotating it and turning it to explore latitude and longitude at the poles and in the Pacific Ocean. Determine the maximum values for latitude and longitude and where those are located.

Exploration 1.3 Location Part B

B. Which of the following coordinate pairs is associated with Lisbon, Portugal?

1. 38°N, 7°W
2. 38°S, 7°W
3. 38°N, 7°E
4. 38°S, 7°E
5. 7°N, 38°W

Exploration 1.3 Location Part C

C. Based on your assessment of latitude and longitude in Google Earth™, which of the following statements is **not** correct?

1. The maximum value of longitude is 90°
2. The North and South Poles are located at 90°N and 90°S.
3. All locations south of the equator have negative values of latitude.
4. All locations located east of the Prime Meridian and west of the International Date Line (Antemeridian) will have east longitude.
5. The equator represents the 0° line of latitude.

Exploration 1.3 Location Part D

D. Based on your assessment of latitude and longitude in Google Earth™, which of the following statements is **not** correct?

1. Lines of latitude are parallel with one another, thus insuring that a degree of latitude is equidistant anywhere on the globe.
2. All of Europe is classified as north latitude.
3. New Zealand's longitudinal values are higher (more eastern) than Australia's.
4. The Prime Meridian represents the 0° line of longitude.
5. Lines of longitude are parallel with one another, thus insuring that a degree of longitude is equidistant anywhere on the globe.

When you complete this exploration, turn off the **1.3 Location** folder.

Exploration 1.3 – Location
Short Answer

Exploration 1.3 Location Short Answer A

A. Use Google Earth™ to determine the latitude and longitude of your university and your home. List the coordinates for both and explain why the numbers are lower or higher in comparison with one another.

_____
_____
_____
_____
_____
_____
_____

Instructions for Exploration 1.3 Location Short Answer B:

1. Turn on and open the **1.3 Location** folder. Double-click the *Mt Kilimanjaro* placemark.

2. Sometimes it might be helpful to exaggerate the variations in elevation in order to see patterns on the landscape. Click Tools and then Options. The Google Earth™ Options

dialog box will open. In the Terrain Quality box change the Elevation Exaggeration to 3 and then click OK.

3. You will need to zoom out to see all of Kilimanjaro. Beyond the mountain being more dramatically represented, notice how the stream channels on the flanks of Kilimanjaro are now more visible. Rotate the scene 360° by dragging the "N" in the compass.

Exploration 1.3 Location Short Answer B

B. Describe the landscape around Mt Kilimanjaro and identify the highest peak within 100 km of Kilimanjaro. What is the latitude and longitude, how tall is it, how far away, and what direction from Kilimanjaro is it?

_____

_____

_____

_____

_____

_____

_____

IMPORTANT: When you've completed this question, return to the Google Earth™ Options dialog box and reset the Elevation Exaggeration to 1.

When you complete this exploration, turn off the **1.3 Location** folder.

Exploration 1.4 – Earth-Sun Relations
Multiple Choice

Google Earth™ is a terrific tool to visualize Earth's complex physical landscapes. The three-dimensional capacity of the program allows the user to get a grasp on the variations in relief that a paper-map is not able to provide.

Instructions for all Parts:

1. Make sure you have opened the **KMZ** file from www.mygeoscienceplace.com

2. From the Places panel, expand **1. Introduction to Google Earth.kmz** then open the **1. Introduction to Google Earth** folder. Open and turn on the **1.4 Earth-Sun Relations** folder.

3. Turn off all layers except the *Borders and Labels* in the *Primary Database*.

4. Turn on the layer depicting the Earth's graticule. To do this, click View and then select Grid.

Instructions for Exploration 1.4 Earth-Sun Relations Parts A-B:

1. Turn on and open the **Latitude Locations** folder.

2. Examine the locations by double-clicking on each of the five placemarks labeled *A* through *E*, paying attention to their location relative to the equator.

Exploration 1.4 Earth-Sun Relations Part A

A. If the date is December 21, at which placemark would the sun be directly overhead at noon?

1. A
2. B
3. C
4. D
5. E

Exploration 1.4 Earth-Sun Relations Part B

B. On June 21st, which location will experience the least amount of sunlight?

1. A
2. B
3. C
4. D
5. E

Turn off the **Latitude Locations** folder.

Instructions for Exploration 1.4 Earth-Sun Relations Parts C–D:

1. Turn on and double-click the **World Timezone Clock** folder.

Exploration 1.4 Earth-Sun Relations Part C

C. Excluding political adjustments to time (such as Daylight Savings Time), which of the following cities would be +8:00 Greenwich Mean Time?

1. Jakarta, Indonesia
2. Urumqi, China
3. Los Angeles, California
4. Dakar, Senegal
5. Suva, Fiji

Exploration 1.4 Earth-Sun Relations Part D

D. Which of the following nations has the highest number of time zones?

1. China
2. Brazil
3. United States
4. Russia
5. Australia

When you complete this exploration, turn off the **1.4 Earth-Sun Relations** folder.

Exploration 1.4 – Earth-Sun Relations Short Answers

Instructions for all Parts:

1. Make sure you have opened the **KMZ** file from the blue box on the left.

2. From the Places panel, expand **1. Introduction to Google Earth.kmz** then open the **1. Introduction to Google Earth** folder. Open and turn on the **1.4 Earth-Sun Relations** folder.

Instructions for Exploration 1.4 Earth-Sun Relations Short Answer A:

1. Turn on and open the **Latitude Locations** folder.

Exploration 1.4 Earth-Sun Relations Short Answer A

A. Choose a date, such as March 22nd. Describe the sun conditions at each of the five placemarks on your selected date.

_____

_____

_____

_____

_____

_____

_____

When you've completed this question, turn off the **Latitude Locations** folder.

The Earth can be divided into 24 equal zones of 15 degrees; however, you may notice that time zones are not neatly divided in this way.

Instructions for Exploration 1.4 Earth-Sun Relations Short Answer B:

1. Turn on and double-click the **World Timezone Clock** folder.

2. Sometimes it might be helpful to exaggerate the variations in elevation in order to see patterns on the landscape. Click Tools and then Options. The Google Earth™ Options dialog box will open. In the Terrain Quality box change the Elevation Exaggeration to 3 and then click OK.

3. You will need to zoom out to see all of Kilimanjaro. Beyond the mountain being more dramatically represented, notice how the stream channels on the flanks of Kilimanjaro are now more visible. Rotate the scene 360° by dragging the "N" in the compass.

Exploration 1.4 Earth-Sun Relations Short Answer B

B. Select a location with a departure from the expected time zone division. Where is this time zone adjustment located? Research this location and determine why the time zone lines do not follow their expected course.

_____

_____

_____

_____

_____

_____

_____

When you complete this exploration, turn off the **1.4 Earth-Sun Relations** folder.

Encounter Physical Geography

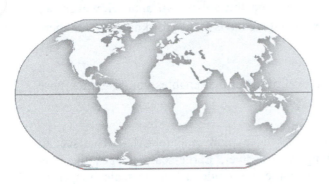

Name:_____

Date: _____

Chapter 2

Portraying Earth

Exploration 2.1 – Scale and Isolines
Multiple Choice

Another important concept to grasp is that of scale. When we use the term scale, we are referring to the relationship between units on a map or, in this case, a digital globe, compared to units in the real world. An example of a verbal scale that you may have seen before would be "1 inch equals 1 mile." A scale can also be portrayed as a ratio, such as 1:1000. This means that one unit on the map is equal to 1000 units in the "real world." Large-scale maps show a large amount of detail but not much area, whereas a small-scale map shows a more extensive area with less detail. Therefore, a 1:1000 scale map is a larger scale map than a 1:5000 scale map. As you will experience in your work with Google Earth™, some types of analysis are better suited to viewing the Earth at larger scales (zoomed in more), while others dictate a small-scale approach (zoomed out more).

Instructions for all Parts:

1. Make sure you have opened the **KMZ** file from www.mygeoscienceplace.com

2. From the Places panel, expand **2. Portraying Earth.kmz** and then open the **2.1 Scale and Isolines** folder.

Instructions for Exploration 2.1 Scale and Isolines Parts A–B

1. Open and turn on the **Denver Scale** folder. Double-click the *Denver 1* placemark. You see the greater Denver metropolitan area. It's difficult to pick out many features of the metropolitan area at this scale. You should be able to identify the mountains to the west of town, some of the stream networks, agricultural regions, and the urban area. This is the smallest scale view of Denver that we will utilize.

2. Double-click the *Denver 2* placemark. This view is a larger-scale view than the one associated with *Denver 1*. By zooming in we can see much more detail, but we've lost some of the big picture. For example, the high mountains are no longer visible. But now we can see highways and the major roads and features such as reservoirs and green space.

3. Move to the *Denver 3* placemark. Much has been gained by zooming in. This is a larger-scale view than *Denver 2*, but will be smaller scale than *Denver 4*. Now we see features such as sports stadiums, highway interchanges, and the downtown area. However, we have no idea how big the metropolitan area is and we have no way of knowing that there are mountains immediately to the west.

4. Double-click *Denver 4* and you will see a level of detail that enables you to gain a more in-depth understanding of the features on the ground.

5. Double-click *Denver 5* and the detail becomes even more apparent. Here we see an amusement park and its parking lot along with a river. Large-scale images like this facilitate analysis and interpretation. For example, we could use large-scale images to determine the spatial distribution of crosswalks or the amount of the park that was shaded by trees.

Exploration 2.1 Scale and Isolines Part A

A. If you found it necessary to have an even smaller scale perspective of Denver than is provided in the *Denver 1* through *Denver 5* placemarks, which of the following actions would you perform?

1. Zoom in from the *Denver 1* placemark.
2. Zoom out from the *Denver 1* placemark.
3. Zoom in from the *Denver 5* placemark.
4. Zoom out from the *Denver 5* placemark.
5. Turn on the *Roads* layer in the *Primary Database* while viewing the *Denver 1* placemark.

Exploration 2.1 Scale and Isolines Part B

B. If you wanted to provide a rough estimate of surface parking available within 500 meters of the Pepsi Center, noted by a placemark of the same name, which of the *Denver* placemarks would provide the best possible scale for your analysis?

1. *Denver 1*
2. *Denver 2*
3. *Denver 3*
4. *Denver 4*
5. *Denver 5*

When you complete these parts, turn off and collapse the **Denver Scale** folder.

Instructions for Exploration 2.1 Scale and Isolines Part C

1. Double-click the *Chief Mountain Quadrangle* overlay. This is a United States Geological Survey (USGS) 1:24000-scale topographic map that is utilized in a variety of industries. The brown lines on the map indicate locations of equal elevation with a contour interval of 40 feet.

2. Examine the map, paying particular attention to the contour gradient across this area. The closer the contour lines, the steeper the slope.

Exploration 2.1 Scale and Isolines Part C

C. Which of the following locations on the *Chief Mountain Quadrangle* overlay has the steepest contour gradient?

1. Southeast face of Sherburne Peak.
2. North face of Yellow Mountain.
3. The north side of Sandy Ridge.
4. The south side of Sandy Ridge.
5. The southwest face of Chief Mountain.

Instructions for Exploration 2.1 Scale and Isolines Part D:

1. Turn on and double-click the **African Rain** folder. This overlay shows the precipitation in Africa on June 12, 2012, with progressively higher amounts of rain indicated by darker shades of green.

2. Click on an area of precipitation in the 3D Viewer to determine the amount of rain that fell at any given location. A pop-up window will display the information stored within this precipitation dataset. For this data, the value for Contour indicates the precipitation in millimeters.

Exploration 2.1 Scale and Isolines Part D

D. Based on the data in the **African Rain** folder, which of the following locations had the highest rainfall on June 12, 2012?

1. Abuja, Nigeria
2. Dakar, Senegal
3. Freetown, Sierra Leone
4. Ouagadougou, Burkina Faso
5. Yaounde, Cameroon

When you complete this exploration, collapse and turn off the **2.1 Scale and Isolines** folder.

Exploration 2.1 – Scale and Isolines
Short Answer

Instructions for all Parts:

1. Make sure you have opened the **KMZ** file from www.mygeoscienceplace.com

2. From the Places panel, expand **2. Portraying Earth.kmz** and then open the **2.1 Scale and Isolines** folder.

Instructions for Exploration 2.1 Scale and Isolines Short Answer A:

1. Open and turn on the **Denver Scale** folder. Examine the relationship between the five placemarks labeled *Denver 1* through *Denver 5*.

Exploration 2.1 Scale and Isolines Short Answer A

A. Identify two basic visual analysis tasks (e.g., measuring the size of parking lots) and explain which of the placemarks would be appropriate to perform those tasks and why.

_____

_____

_____

_____

_____

_____

_____

Instructions for Exploration 2.1 Scale and Isolines Short Answer B:

1. Double-click the *Chief Mountain Quadrangle* overlay. This is a United States Geological Survey (USGS) 1:24000-scale topographic map that is utilized in a variety of industries. The brown lines on the map indicate locations of equal elevation with a contour interval of 40 feet.

2. Utilize the tilt capabilities of Google Earth™ by employing the Look joystick in the Navigation panel. This will allow you to see a new perspective of the map.

3. Adjust the transparency of this layer by using the Adjust Opacity button at the bottom of the Places panel. Use the slider to change the transparency level.

Exploration 2.1 Scale and Isolines Short Answer B

B. How does changing the tilt of the view or the opacity of the topographic map overlay of Chief Mountain increase or decrease your ability to understand the landscape? What features are easier to identify with a tilted view or an opaque topographic map overlay?

_____

_____

_____

_____

_____

_____

_____

When you complete this exploration, collapse and turn off the **2.1 Scale and Isolines** folder.

Exploration 2.2 – Remotely Sensed Data
Multiple Choice

Google Earth™ makes extensive use of remotely sensed imagery. This exploration will help you understand some of the issues you are likely to encounter when working with remotely sensed imagery in Google Earth™.

Instructions for all Questions

1. Make sure you have opened the **KMZ** file from www.mygeoscienceplace.com

2. From the Places panel, expand **2. Portraying Earth.kmz** and then open the **2.2 Remotely Sensed Data** folder.

Instructions for Exploration 2.2 Remotely Sensed Data Part A:

1. Double-click the *Resolution 1* placemark. When we use the term resolution we are referring to the spatial resolution or the measurement of the minimum distance between two objects that will allow them to be differentiated from one another in an image. A higher resolution image would have a lower number associated with it. For example, a 30-meter spatial resolution image would be a higher resolution image than a 100-meter spatial resolution image. The imagery to the right of the *Resolution 1* placemark has a higher spatial resolution than the imagery to the left. Notice how individual features, such as trees, are easier to discern on the right.

2. Zoom in and see how you can see the intricacies of the shoreline in the body of water on the right compared to the image on the left where features are much more generalized.

3. Turn on and open the **Resolution Assessment** folder. Examine the placemarks *A* through *E* individually, zooming in to a consistent eye altitude at each (e.g., 2 kilometers).

Exploration 2.2 Remotely Sensed Data Part A

A. Which of the following placemarks in the **Resolution Assessment** folder is located in the area with the lowest spatial resolution imagery?

1. A
2. B
3. C
4. D
5. E

Instructions for Exploration 2.2 Remotely Sensed Data Part B:

1. Double-click the *Imagery Sets* placemark. The varied imagery (difference in color and resolution) of adjacent locations may be caused by the fact that images have been captured by different sensors, have been taken at different times of the year, or were captured in different years.

You can usually find out the date an image was captured by zooming into an eye altitude below 5 kilometers. The imagery date(s) will appear in the bottom left-hand corner of the screen. Be wary of the fact that Google will sometimes add color to imagery to help smooth the edges of image sets.

Exploration 2.2 Remotely Sensed Data Part B

B. How many imagery sets does the view associated with the *Imagery Sets* placemark contain?

1. 1
2. 2
3. 3
4. 4
5. greater than 4

It is important for you to realize that you are usually not limited to viewing an area of the Earth at only one snapshot in time. The historical imagery function of Google Earth™ can be very helpful in highlighting change that takes place over time.

Instructions for Exploration 2.2 Remotely Sensed Data Part C:

1. In the Google Earth™ toolbar, the Historical Imagery button has a picture of a clock with an arrow across the top. If the tool is on, the button will be selected and a time slider will appear in the upper left-hand corner of the 3D Viewer. Ensure that the Historical Imagery tool is turned on. Use this slider to see the landscape at various points in time.

2. Turn on and double-click the *Imagery Through Time* placemark. You will zoom to an area near Birmingham, Alabama. Three placemarks labeled *Site A*, *Site B*, and *Site C* should be visible. The variations in imagery captured in different seasons can also impact your ability to interpret landscapes. For example, significant urban forests can obstruct the views of communities from above.

Exploration 2.2 Remotely Sensed Data Part C

C. Which of the following statements is **not** supported by the historical Google Earth™ imagery?

1. High resolution imagery for the area is available in Google Earth™ dating back to 1978.
2. Leaf-on and leaf-off imagery is available in the historical series.
3. Buildings are constructed at the *Site A* placemark some time between 1998 and 2002.
4. *Site B* has forest cleared between 2005 and 2007.
5. Roads are constructed at *Site C* between 2005 and 2006.

Instructions for Exploration 2.2 Remotely Sensed Data Part D:

1. Turn on and double-click the *Image Variation by Date* placemark. This is a case where you have two images of different vintages abutting one another. Evaluate the scene, using the Historical Imagery slider as necessary, and think about what kinds of problems could arise because of this.

Exploration 2.2 Remotely Sensed Data Part D

D. Which of the following evaluations of Heathrow Airport (*Image Variation by Date* placemark) could be completed with the available imagery?

    1. Total number of aircraft on the ground on September 27, 2008.
    2. Total number of aircraft on the ground on March 4, 2008.
    3. Total number of aircraft on the ground on June 26, 2010.
    4. Length of Heathrow's northernmost runway on June 26, 2010.
    5. Total number of passengers at the airport on September 27, 2008.

IMPORTANT: Be sure to turn off the Historical Imagery tool.

When you complete this exploration, collapse and turn off the **2.2 Remotely Sensed Data** folder.

Exploration 2.2 – Remotely Sensed Data
Short Answer

Exploration 2.2 Remotely Sensed Data Short Answer A

A. What types of problems could arise when interpreting a scene that contains imagery from different years? Specifically, provide some examples of changes that can occur in both the physical and cultural landscape over time.

_____

_____

_____

_____

_____

_____

_____

Instructions for Exploration 2.2 Remotely Sensed Data Short Answer B:

    1. Make sure you have opened the **KMZ** file from www.mygeoscienceplace.com

    2. From the Places panel, expand **2. Portraying Earth.kmz** and then open the **2.2 Remotely Sensed Data** folder.

    3. Ensure that the Historical Imagery tool is turned on. If it is not, click on the Historical Imagery button on the tool bar.

    4. Double-click the *Imagery Through Time* placemark.

Exploration 2.2 Remotely Sensed Data Short Answer B

B. What are the most noticeable changes you can identify in the vicinity of the *Imagery Through Time* placemark? How would you describe development in this region? Is it happening rapidly, moderately, or not at all? Support your answer with examples from the historical imagery as well as the default Google Earth™ imagery, which can be viewed by turning off the Historical Imagery tool.

_____

_____

_____

_____

_____

_____

_____

IMPORTANT: Be sure to turn off the Historical Imagery tool.
When you complete this exploration, collapse and turn off the **2.2 Remotely Sensed Data** folder.

Exploration 2.3 – Image Interpretation I
Multiple Choice

Throughout the text you will be asked to problem-solve by identifying unique features or providing some type of interpretive analysis of physical landscapes. There are several elements of interpretation that are usually present to some degree in any imagery. Exploration 2.3 examines elements such as color, size, shape, and texture.

Instructions for all Parts:

1. Make sure you have opened the **KMZ** file from www.mygeoscienceplace.com

2. From the Places panel, expand **2. Portraying Earth.kmz** and then open the **2.3 Image Interpretation I** folder.

Tone and/or color is one of the most distinguishing characteristics of a given feature in Google Earth™.

Instructions for Exploration 2.3 Image Interpretation I Part A:

1. Double-click the *Tone/color 1* placemark and you will see a large football stadium. The lush green of the pitch stands out. Another field might not be so well-maintained and would not be the deep, even green we see at Camp Nou, home to FC Barcelona. We also see the distinctive red roofs of the region, the gray of asphalt, and the numerous colors of automobiles.

2. Double-click the *Tone/color 2* and view the coastal waters. The varied shades of blues and greens can indicate the presence of subsurface vegetation or coral structures. Here, the darker colors are related to deeper channels.

3. Double-click the *Tone/color 3* placemark and examine the landscape for variations in tone or color.

Exploration 2.3 Image Interpretation I Part A

A. Based on the tone and/or color clues at the *Tone/color 3* placemark, what is the most likely surface cover?

1. cotton fields
2. forest
3. sand
4. snow and ice
5. water

Size can help your interpretation in relative and absolute ways.

Instructions for Exploration 2.3 Image Interpretation I Part B:

1. Double-click the *Size 1* placemark. You will see Boone Pickens Stadium at Oklahoma State University. We can obtain an absolute size from this image because we know that an American football field is 120 yards (about 110 meters) long if you include the end zones. You can apply this knowledge to measure the structure next to the stadium, Gallagher-Iba Arena. We also see opportunities for relative size evaluation. Automobiles vary in size somewhat, but they can provide a comparative reference point.

2. Double-click the *Size 2* placemark to see another really good example of an absolute size marker. The vast majority of railroad tracks in the United States are a standardized gauge with the rails placed 4' 8.5" (1.48 meters) apart from one another.

3. Double-click the *Size 3* placemark and survey the landscape looking for features that could assist in length or area measurements.

Exploration 2.3 Image Interpretation I Part B

B. If you wanted to determine the absolute size of the reservoir at the *Size 3* placemark, what feature visible in the scene would be the best choice to use as your size reference?

1. the width of a road
2. a parking lot
3. roof area of a house
4. tree crown size
5. a tennis court

Instructions for Exploration 2.3 Image Interpretation I Part C:

1. Double-click the *Shape 1* placemark. You will fly to an industrial agriculture landscape. These are center-pivot irrigation systems on the Great Plains. They utilize groundwater pumped to the surface and distributed by a sprinkler that rotates around a fixed pivot in the center of the field.

2. Double-click the *Shape 2* placemark. You may recognize this feature as it is named after its shape. The Pentagon in Washington, DC, is one of the world's largest structures.

3. Double-click the *Shape 3* placemark. Explore this scene, paying particular attention to the distinctive linear shapes.

Exploration 2.3 Image Interpretation I Part C

C. What are the distinctive linear shapes seen at the *Shape 3* placemark?

1. airport runways
2. interstate highways
3. streets in a housing development that has yet to be built
4. skyscrapers
5. railroad tracks

The texture of a surface can be quite fine or very coarse. This can give the viewer an idea about the types of vegetation on the ground or about the degree of homogeneity of a landscape.

Instructions for Exploration 2.3 Image Interpretation I Part D:

1. Double-click *Texture 1* and view the dense forest. This is a relatively uniform landscape.

2. Double-click *Texture 2*. This scene, on the other hand, has a significant degree of variation. You are viewing an ecotone as the landscape transitions from steppe into woodlands and forest from east to west. This is a result of changing elevation and precipitation.

3. Double-click the *Texture 3* placemark and evaluate the landscape at that location.

Exploration 2.3 Image Interpretation I Part D

D. Which of the following statements is best supported by the evidence at the *Texture 3* placemark?

1. This is a landscape with no variation in texture.
2. This landscape has significant variation in texture as a result of some agricultural activity.
3. Texture varies in a regular and geometric pattern.
4. The western side of this scene has much more variation in texture than the eastern side.
5. This landscape has high texture variation because of its combination of vegetation and the built environment.

When you complete this exploration, collapse and turn off the **2.3 Image Interpretation I** folder.

Exploration 2.3 – Image Interpretation I
Short Answer

Instructions for Exploration 2.3 Image Interpretation I Short Answer A:

1. Zoom in to your neighborhood or the neighborhood around your school.

2. Identify a feature based on at least two elements of image interpretation.

Exploration 2.3 Image Interpretation I Short Answer A

A. Note the coordinates (latitude and longitude) of this feature in addition to providing an interpretation.

_____

_____

_____

_____

_____

_____

_____

Some image overlays you will utilize in this text will incorporate multispectral image interpretation. These images are captured using electromagnetic spectrum bands outside the range of visible light. Examples include the use of infrared or thermal imagery.

Instructions for Exploration 2.3 Image Interpretation I Short Answer B:

1. Make sure you have opened the **KMZ** file from the blue box on the left.

2. From the Places panel, expand **2. Portraying Earth.kmz** and then open the **2.3 Image Interpretation I** folder.

3. Turn on and double-click the *Urban Heat Island* overlay. This image was collected by the Landsat 7 satellite Enhanced Thematic Mapper system and shows the heat energy detected in the Buffalo, New York, area with darker areas representing relatively cool zones and brighter areas representing hotter zones.

4. Turn the image on and off and compare it to the features in the default Google Earth™ imagery.

Exploration 2.3 Image Interpretation I Short Answer B

B. Explain the relationship between urban areas, suburban areas, and agricultural areas and recorded temperatures.

_____

_____

_____

_____

_____

_____

_____

When you complete this exploration, collapse and turn off the **2.3 Image Interpretation I** folder.

Exploration 2.4 – Image Interpretation II
Multiple Choice

Continuing with the theme of image interpretation in Google Earth ™, Exploration 2.4 examines elements such as pattern, association, shadow, and site/situation.

Instructions for all Parts:

1. Make sure you have opened the **KMZ** file from the blue box on the left.

2. From the Places panel, expand **2. Portraying Earth**.kmz and then open the **2.4 Image Interpretation II** folder.

More often than not, regular patterns on the landscape suggest human involvement.

Instructions for Exploration 2.4 Image Interpretation II Part A:

1. Double-click the *Pattern 1* placemark. This location shows natural vegetation that is being intensively managed by humans. In this case, we are looking at one of the world's largest pecan orchards.

2. Double-click the *Pattern 2* placemark. The concentric rings shown here are associated with surface mining. As this mining operation continues, this pattern will increase in depth and width (at the expense of the existing community).

3. Double-click one of the *Pattern 3* placemarks. This is a planned suburb in the western United States where the houses are all very similar in size and appearance.

4. Zoom in to one of the *Pattern 3* placemarks and examine the details of this structure.

Exploration 2.4 Image Interpretation II Part A

A. What are the buildings that are indicated by the *Pattern 3* placemarks?

1. homes for the very wealthy
2. retail establishments
3. schools
4. factories
5. hospitals

Association refers to the ability to identify or confirm the existence of a feature based on its relationship to other features.

Instructions for Exploration 2.4 Image Interpretation II Part B:

1. Double-click the *Association 1* placemark. You will see a collection of long yellow vehicles in a parking lot. These are school buses. Why would you have that many school buses in one location? Perhaps this is the bus yard for a large metropolitan school district?

2. Zoom out and you will see that it is not a large city, but there is a large industrial building adjacent to the buses. Research would reveal that this is Fort Valley, Georgia, home of the Bluebird Bus Company factory.

3. Double-click the *Association 2* placemark. This location contains strange groupings of what appear to be parts of airplanes.

4. Zoom out to get a better view of the entire area. You will see that there are numerous intact airplanes and also an airport nearby, which explains how this material has arrived. This is one of the US military's airplane bone yards, which are usually found in desert areas because the planes do not rust as quickly in the dry climate.

5. Double-click the *Association 3* placemark. Examine the landscape, paying particular attention to the context of this feature.

Exploration 2.4 Image Interpretation II Part B

B. What is the feature seen at the *Association 3* placemark?

1. a railroad bridge
2. a dam
3. a combo rail and highway bridge
4. an aqueduct
5. a highway bridge

After completing this part, be sure to turn off the *Photos* layer in the *Primary Database*.

Shadows can be helpful in providing an idea of a feature's height and shape, as it can sometimes be difficult to ascertain this from a vertical perspective above the object.

Instructions for Exploration 2.4 Image Interpretation II Part C:

1. Double-click the *Shadow 1* placemark. This placemark shows the core of the Houston, Texas, metropolitan area. Downtown Houston is distinguishable based on its buildings' shadows. Excluding a few tall buildings due west of downtown there is little vertical development of real estate outside the central business district.

2. Double-click the *Shadow 2* placemark. The distinctive shadow at this location is associated with a cooling tower at one of France's largest nuclear power plants. Notice that the cooling tower is grainier than the rest of the image. Also, where is the shadow for the second cooling tower? This imagery may have been modified or distorted for security reasons.

3. Double-click the *Shadow 3* placemark. You should see a view of Washington, DC.

4. Check the imagery date at the bottom left of the view and ensure that the date is 05/18/2008; if it is not, utilize the Historical Imagery tool.

5. Examine the landscape at *Shadow 3*, paying particular attention to the shadows cast by the buildings.

Exploration 2.4 Image Interpretation II Part C

C. Based on the evidence in the *Shadow 3* placemark, what time of day was this image was captured?

1. 8 a.m.
2. noon
3. 2 p.m.
4. 5 p.m.
5. 8 p.m.

IMPORTANT: After completing this part, turn off the Historical Imagery tool.

Site/situation refers to the geographic context of the feature, features, or landscape that you are interpreting. Site refers to the local circumstances; situation refers to the regional context.

Instructions for Exploration 2.4 Image Interpretation II Part D:

1. Double-click the *Site/situation 1* placemark. You will see an airport with a number of planes and helicopters positioned on the tarmac. You will also see a large number of modular structures around the specific location (the site).

2. Zoom out from the *Site/situation 1* placemark.

3. Turn on the *Borders and Labels* layer in the *Primary Database*. It should now be apparent that we are in Afghanistan. This is a military base supporting the war efforts there. The situation of this image explains the aircraft and structures.

4. Double-click the *Site/situation 2* placemark. This location shows the Icelandic volcano that shut down air traffic over much of Europe in April 2010. At the time this text was written, Google did not have updated imagery illustrating the eruption. As you look at this placemark, it is possible that has changed. The moral of the story is: Google Earth™ imagery is not live. Sometimes you will find imagery that is very recent, while other times the imagery may be more than five years old.

5. Double-click the *Site/situation 3* placemark. The details at this location underscore the necessity to be aware of contemporary contexts of Google Earth™ imagery. Examine this scene, which displays the grounds of a presidential palace. Pay particular attention to the presence of thousands temporary tents in what should be a relatively formal location.

Exploration 2.4 Image Interpretation II Part D

D. Based on the evidence at the *Site/Situation 3* placemark and the context of the image, what event has recently occurred at this location?

1. an earthquake
2. Earth Day
3. a massive demonstration against the government
4. a tsunami
5. a street festival

IMPORTANT: When you complete this part, turn off the *Photos* layer in the *Primary Database* and turn off the Historical Imagery tool.

When you complete this exploration, collapse and turn off the **2.4 Image Interpretation II** folder.

Exploration 2.4 – Image Interpretation
Short Answer

Instructions for Exploration 2.4 Image Interpretation II Short Answer A:

1. Make sure you have opened the **KMZ** file from www.mygeoscienceplace.com

2. From the Places panel, expand **2. Portraying Earth.kmz** and then open and turn on both the **2.3 Image Interpretation I** and **2.3 Image Interpretation I** folders.

Exploration 2.4 Image Interpretation II Short Answer A

A. Select one of the interpretation placemarks from the **2.3 Image Interpretation I** or **2.4 Image Interpretation II** folders and describe the evidence of all of the interpretive elements visible in that scene.

_____

_____

_____

_____

_____

_____

_____

_____

Exploration 2.4 Image Interpretation II Short Answer B

B. What two elements of interpretation do you think are most helpful to you when trying to understand Google Earth™ imagery? Explain your answers and provide examples.

_____

_____

_____

_____

_____

_____

_____

When you complete this exploration, collapse and turn off the **2.4 Image Interpretation II** folder.

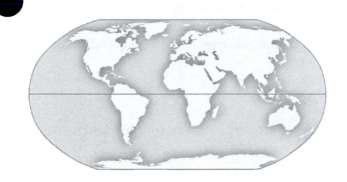

Encounter Physical Geography

Name:_____

Date: _____

Chapter 3

Introduction to the Atmosphere

Exploration 3.1 – Atmospheric Profile
Multiple Choice

Most of our attention is focused on the troposphere but it is important to develop a better understanding of the profile and composition of the entire atmosphere. Processes ranging from weather creation, ultraviolet radiation protection, and radio wave refraction occur at various altitudes in the atmosphere.

Instructions for all Parts:

1. Make sure you have opened the **KMZ** file from www.mygeoscienceplace.com

2. From the Places panel, expand **3. Introduction to the Atmosphere.kmz** and then open the **3.1 Atmospheric Profile** folder.

Instructions for Exploration 3.1 Atmospheric Profile Part A:

1. Turn on and double-click the **Profile** folder. The column shown represents a generalized cross-section of the atmosphere.

2. Turn on and open the **Layers** folder. Examine each of the placemarks for layers of the atmosphere located next to a corresponding profile section.

Exploration 3.1 Atmospheric Profile Part A

A. Which labeled placemark in the **Layers** folder represents the stratosphere?

1. A
2. B
3. C
4. D
5. E

When you complete this part, collapse and turn off the **Layers** folder.

Instructions for Exploration 3.1 Atmospheric Profile Part B:

1. Ensure the **Profile** folder is on then turn on and open the **Altitude** folder.

2. Examine the five placemarks based on their altitude and relative location compared to the atmospheric profile.

Exploration 3.1 Atmospheric Profile Part B

B. Which of the labeled placemarks in the **Altitude** folder represents the approximate location of aurora events?

1. A
2. B
3. C
4. D
5. E

When you complete this part, collapse and turn off the **Altitude** folder.

Instructions for Exploration 3.1 Atmospheric Profile Part C:

1. Ensure the **Profile** folder is on, then turn on and open the **Composition** folder.

2. Double-click on each of the placemarks labeled *A* through *E* individually. Note the altitude and relative location in the atmosphere of each.

Exploration 3.1 Atmospheric Profile Part C

C. Based on the altitude of each placemark in the **Composition** folder, which of the placemarks would be in a location where atmospheric gases are mixed together?

1. A
2. B
3. C
4. D
5. E

When you complete this part, turn off the **Profile** folder and collapse and turn off the **Composition** folder.

Instructions for Exploration 3.1 Atmospheric Profile Part D:

1. Turn on and double-click the **Temperature** folder. Examine each of the five placemarks labeled *A* through *E* individually, paying particular attention to their altitude relative to the thermal layers of the atmosphere. The Eye Altitude for each placemark is noted on the right side of the Status Bar at the bottom of the 3D Viewer.

Exploration 3.1 Atmospheric Profile Part D

D. At which of the placemarked locations in the **Temperature** folder would the average temperature be the lowest?

1. A
2. B
3. C
4. D
5. E

When you complete this exploration, turn off and collapse the **3.1 Atmospheric Profile** folder.

Exploration 3.1 – Atmospheric Profile
Short Answer

Instructions for all Parts:

1. Make sure you have opened the **KMZ** file from www.mygeoscienceplace.com

2. From the Places panel, expand **3. Introduction to the Atmosphere.kmz** and then open the **3.1 Atmospheric Profile** folder.

Instructions for Exploration 3.1 Atmospheric Profile Short Answer A:

1. Open the **Altitude** folder and turn on and double-click placemark *C*. In 1960, US Air Force Captain Joseph Kittinger leapt from a balloon from this altitude (32,000 meters) as part of a government testing project named Excelsior. The tests were meant to evaluate equipment designed for the nascent manned space program and focused on the extreme conditions found in the atmosphere and beyond.

Exploration 3.1 Atmospheric Profile Short Answer A

A. Using the information in your textbook, describe the conditions Captain Kittinger would have experienced on his descent from the altitude at placemark *C*. Be certain to indicate the temperature and pressure changes that would have occurred at various altitudes.

_____
_____
_____
_____
_____

Instructions for Exploration 3.1 Atmospheric Profile Short Answer B:

1. Turn on and open the **Comparison** folder.

2. Double-click *Placemark A* and *Placemark B* individually. Note the location for each, paying attention to latitude and longitude and relative location on the globe.

Exploration 3.1 Atmospheric Profile Short Answer B

B. Describe the expected difference in the depth of the troposphere between these two locations. Be certain to indicate why these differences are present.

_____

_____

_____

_____

_____

When you complete this exploration, turn off and collapse the **3.1 Atmospheric Profile** folder.

Exploration 3.2 – Ozone
Multiple Choice

Ozone at the surface is problematic for human health, but ozone in the stratosphere is important for the health of all biota on the planet. Unfortunately, anthropogenic chemicals have dramatically thinned the Earth's protective ozone layer. This exploration will examine the changes in this important atmospheric compound.

Instructions for all Parts:

1. Make sure you have opened the **KMZ** file from www.mygeoscienceplace.com

2. From the Places panel, expand **3. Introduction to the Atmosphere.kmz** and then open the **3.2 Ozone** folder.

Instructions for Exploration 3.2 Ozone Part A:

1. Turn on and open the **Ozone Hole** folder.

2. Double-click the _Ozone Hole Image_ layer to fly to and view a recent representation of the ozone hole. Stratospheric ozone depletion is particularly problematic for one region of Earth but this image is not appropriately centered on that region.

3. View each of the placemarks contained in the **Ozone Hole** folder individually, paying attention to their location relative the area shown in the _Ozone Hole Image_ layer.

Exploration 3.2 Ozone Part A

A. Which of the placemarks in the **Ozone Hole** folder is the most appropriate location for the _Ozone Hole Image_?

1. A
2. B
3. C
4. D
5. E

Instructions for Exploration 3.2 Ozone Part B:

1. Ensure that the **Ozone Hole** folder is open. If any layers are turned on inside this folder, turn them off.

2. Click the hyperlink under the *Ozone Hole Image* layer to open NASA's Ozone Hole Watch webpage. Here you can find information and historic data about stratospheric ozone depletion.

Exploration 3.2 Ozone Part B

B. Based on the information presented on the front page of Ozone Hole Watch, which of the following statements is most strongly supported?

1. Most of the imagery is focused on the polar Arctic region.
2. The ozone hole area reached its largest extent in March of 2008.
3. The average ozone hole area has shown signs of stabilization or a small reduction since the mid-2000s.
4. Ozone is a blue-green gas.
5. The measurement unit for total ozone is known as an ozonometer.

Instructions for Exploration 3.2 Ozone Part C:

1. Ensure that the **Ozone Hole** folder is open.

2. Click the hyperlink under the *Ozone Hole Image* layer to open NASA's Ozone Hole Watch webpage. Here you can find information and historic data about stratospheric ozone depletion. We can use the historical imagery capabilities of the Ozone Hole Watch page to view data for prior months.

3. Click the Ozone Maps tab and scroll down to the Map archives. The linked months provide imagery of average conditions of the ozone hole for that point in time.

4. Click the link for September 2006 and then click the image for the 21st day of the month. Examine the areas of low ozone concentration, paying attention to their location relative to the Earth's latitude and longitude grid system. If you need to, return to Google Earth™ to verify the latitude and longitude of the low ozone area.

Exploration 3.2 Ozone Part C

C. Based on the September 21, 2006, imagery from the Ozone Hole Watch webpage, which of the following locations would have been under the region with the lowest concentration of stratospheric ozone?

1. 60° S, 20° W
2. 90° N, 15° E
3. 44° N, 31° W
4. 75° N, 25° E
5. 25° S, 70° E

As inhabitants of Earth, we are concerned with exposure to the sun's ultraviolet radiation. The ozone layer is an important protective mechanism for the globe. However, some areas are naturally inclined to higher levels of UV radiation due to their latitudinal location and their predominant meteorological conditions.

Instructions for Exploration 3.2 Ozone Part D:

1. Ensure that the Google Earth™ 3D Viewer is visible.

2. Open and turn on the **Historic UV Index** folder.

3. Evaluate the location of the five placemarked cities, noting their locations relative to the equator.

4. Click the hyperlink under the Historic UV Index heading to open the Climate Prediction Center's UV Index page. You will see links to view the UV index for major cities in each state.

5. Using both the latitude of each city and the data available on the UV Index page, determine the relative number of "extreme" UV days experienced at each location.

Exploration 3.2 Ozone Part D

D. Which of the cities placemarked in the **Historic UV Index** folder had the greatest number of days in the UV Index's "Extreme" category in 2011?

1. Billings, Montana
2. Bismarck, North Dakota
3. Chicago, Illinois
4. Indianapolis, Indiana
5. Miami, Florida

When you complete this exploration, turn off and collapse the **3.2 Ozone** folder.

Exploration 3.2 – Ozone
Short Answers

Instructions for all Parts:

1. Make sure you have opened the **KMZ** file from www.mygeoscienceplace.com

2. From the Places panel, expand **3. Introduction to the Atmosphere.kmz** and then open the **3.2 Ozone** folder.

Instructions for Exploration 3.2 Ozone Short Answer A:

1. Click the hyperlink under the **Historic UV Index** folder heading to open the Climate Prediction Center's UV Index webpage. You will see links to view the UV index for major cities in each state.

2. Examine the data from the city located nearest to your location.

Exploration 3.2 Ozone Short Answer A

A. What city is the closest to your location? In what months are the highest UV indices typically recorded? What strategies should you utilize to mitigate the impacts of the high UV values during these times?

_____

_____

_____

_____

_____

Instructions for Exploration 3.2 Ozone Short Answer B:

1. Click the hyperlink under the **Today's UV Index** folder heading to access current information about the UV index in the United States. Scroll down on the page to see a map of the current UV index across the country.

2. Examine this UV index forecast and click the buttons to see the UV forecast change over the next four days. Note the status of the UV index at your location.

Exploration 3.2 Ozone Short Answer B

B. What factors are enhancing or diminishing the UV index at your location over the forecast period? Be sure to identify your location and the date range of your analysis in your response.

_____

_____

_____

_____

_____

When you complete this exploration, turn off and collapse the **3.2 Ozone** folder.

Exploration 3.3 – Air Pollution
Multiple Choice

Beyond the degradation of stratospheric ozone, humans have contributed a wide range of pollutants to the Earth's atmosphere. Impacts include decreased visibility, an increase in respiratory ailments, and the warming of the Earth's atmosphere.

Instructions for all Parts:

1. Make sure you have opened the **KMZ** file from www.mygeoscienceplace.com

2. From the Places panel, expand **3. Introduction to the Atmosphere.kmz** and then open the **3.3 Air Pollution** folder.

Instructions for Exploration 3.3 Air Pollution Part A:

1. Open and turn on the **Inversion** folder.

2. Fly to the *Inversion 1* and *Inversion 2* layers individually. Each location shows active inversions captured by MODIS satellite imagery. Note that the topography of each location may have an impact on the severity or extent of temperature inversion events.

Exploration 3.3 Air Pollution Part A

A. Which of the two explanations below best explains the situations that exist at *Inversion 1* and *Inversion 2*, respectively?

1. 1 – Pollution trapped in relatively lower river basins; 2 – Pollution blocked by a large orographic barrier.
2. 1 – Pollution blocked by a large orographic barrier; 2 – Pollution trapped in relatively lower river basins.
3. Inversions are occurring at both locations due to a concentration of pollution in Polar Zones.
4. 1 – Inversion is located here due to a high population density; 2 – Pollution trapped in relatively lower river basins.
5. 1 – Pollution blocked by a large orographic barrier; 2 – This inversion has been generated by activities in Korea.

When you complete this question, collapse and turn off the **Inversion** folder.

The introduction of particulate matter into the atmosphere can occur in a variety of ways. In some instances, particulates are introduced through natural phenomena, while in others they are introduced via human, or anthropogenic, activities.

Instructions for Exploration 3.3 Air Pollution Part B:

1. Turn on and open the **Particulate Matter** folder.

2. Examine the landscapes at each of the placemarks labeled *A* through *E* individually. Note any evidence of particulate matter being introduced into the atmosphere.

Exploration 3.3 Air Pollution Part B

B. Which of the placemarks in the **Particulate Matter** folder does **not** show the potential release of particulate matter into the atmosphere?

1. A
2. B
3. C
4. D
5. E

When you complete this question, collapse and turn off the **Particular Matter** folder.

Instructions for Exploration 3.3 Air Pollution Part C:

1. Open the **Pollution Emissions** folder.

2. Turn on and double-click the **Total CO$_2$ Emissions** folder to view a map of the world showing the total CO$_2$ emissions including land use change for the year 2000. The data have been extruded to show relative levels of CO$_2$ emissions for each country.
   You can click on a country to open a pop-up box showing that nation's CO$_2$ emissions and the values for the highest and lowest emission nations. These data are a legacy dataset from the World Resources Institute; the original source website may not provide these data.

3. Rotate your view of the globe to see the relative emissions of each country and utilize the pop-up boxes to determine the nations that ranked third and fourth in total CO$_2$ emissions in 2000.

Exploration 3.3 Air Pollution Part C

C. What factor is disproportionally responsible for the total CO$_2$ emissions in the nations ranked 3rd and 4th in 2000?

1. These countries are amongst the most affluent nations in the world.
2. These countries rely heavily on the burning of fossil fuels for heating.
3. Both countries have high volcanic activity.
4. These countries rely heavily on the burning of fossil fuels for electricity.
5. Both countries have a high amount of active deforestation.

When you complete this part, turn off the **Total CO$_2$ Emissions** folder.

Instructions for Exploration 3.3 Air Pollution Part D:

1. Ensure that the **Pollution Emissions** folder is open.

2. Turn on and double-click the **CO$_2$ Emissions Per Capita** folder. This map displays the emission of CO$_2$ in 2004 normalized by the population of each country with the data extruded from the surface to show relative levels. These data are a legacy dataset from the World Resources Institute; the original source website may not provide these data.

3. Rotate your view of the globe to see the relative CO$_2$ emission per capita of each country. Pay particular attention to regional concentrations or patterns.

Exploration 3.3 Air Pollution Part D

D. Which of the statements below is correct based on the data presented in the map?

1. A global north-south divide is noticeable with the Southern Hemisphere nations contributing more CO_2 on a per capita basis.
2. There are no countries in the Southern Hemisphere in the top ten of per capita CO_2 emissions.
3. The globe's focal point for hydrocarbon production is a region with generally higher CO_2 emissions per capita.
4. In 2004, the United States had the highest CO_2 emissions per capita in the world.
5. Countries with the highest populations produce the highest amount of CO_2 emissions per capita.

When you complete this exploration, collapse and turn off the **3.3 Air Pollution** folder.

Exploration 3.3 – Air Pollution
Short Answers

Instructions for all Parts:

1. Make sure you have opened the **KMZ** file from www.mygeoscienceplace.com

2. From the Places panel, expand **3. Introduction to the Atmosphere.kmz** and then open the **3.3 Air Pollution** folder.

Topography plays a key role in establishing the conditions for a temperature inversion to occur.

Instructions for Exploration 3.3 Air Pollution Short Answer A:

1. Double-click the *Inversion Region* placemark to view the Los Angeles Basin. This is a region that often experiences temperature inversions. We can use Google Earth's™ capabilities to better understand the topographic characteristics of the Los Angeles Basin.

2. From the menu bar, click Tools > Options to open the Google Earth™ Options dialogue box. On the 3D View tab, change the Elevation Exaggeration from 1 to 3 and click OK.

Exploration 3.3 Air Pollution Short Answer A

A. Describe the topographic situation of the Los Angeles Basin.

_____
_____
_____
_____
_____
_____
_____

IMPORTANT: When you complete this part, be sure to return to the Google Earth™ Options dialogue and return the Elevation Exaggeration to 1.

Instructions for Exploration 3.3 Air Pollution Short Answer B:

1. Open the **Air Quality** folder. The three cities labeled in the **Air Quality** folder were identified in 2011 as the most polluted US cities in terms of particulate matter. If you were to check the current status of air quality in these cities you would likely find that it was rated less than good. Air quality, however, can vary day-to-day or season-to-season.

2. Click the *Airnow.gov* hyperlink in the **Air Quality** folder to open the website of the government's air quality monitoring program.

3. Use the information on this site to assess today's air quality forecast and identify the three cities with the lowest forecasted level of air quality.

Exploration 3.3 Air Pollution Short Answer B

B. Identify the name of the three cities with the lowest forecasted air quality levels for today. What is the Air Quality Index Level of Health Concern for each? Be sure to note the date this data were retrieved from the AirNow website.

_____
_____
_____
_____
_____

When you complete this exploration, collapse and turn off the **3.3 Air Pollution** folder.

3.4 – Controls of Weather and Climate
Multiple Choice

Short-term and long-term climate patterns are affected by a series of geographic variables. This exploration will examine local and regional climate variation due to controls such as latitude, proximity to water, and altitude.

Instructions for all Parts:

1. Make sure you have opened the **KMZ** file from www.mygeoscienceplace.com

2. From the Places panel, expand **3. Introduction to the Atmosphere.kmz** and then open the **3.4 Controls of Weather and Climate** folder.

Instructions for Exploration 3.4 Controls of Weather and Climate Parts A–B:

1. Turn on and open the **Cities** folder.

2. Examine each of the cities marked by placemarks *A* through *E* individually. Pay particular attention to factors that may influence the local climate such as the presence of nearby large, temperature-moderating water bodies or the city's position relative to the equator.

Exploration 3.4 Controls of Weather and Climate Part A

A. Based on their locations relative to land and water, which of the five cities in the **Cities** folder would likely have the greatest temperature range from the warmest to coldest months?

1. A
2. B
3. C
4. D
5. E

Exploration 3.4 Controls of Weather and Climate Part B

B. Based on their latitudinal locations, which of the five cities in the **Cities** folder would have the lowest temperature during the month of July?

1. A
2. B
3. C
4. D
5. E

When you complete the two previous parts, collapse and turn off the **Cities** folder.

Instructions for Exploration 3.4 Controls of Weather and Climate Part C:

1. Turn on and open the **Altitude** folder. Altitude can have a substantial impact on the temperature range and variation at particular sites.

2. Examine each of the cities marked by placemarks *A* through *E* individually. Use the data display in Google Earth™ to evaluate the elevation at each location.

Exploration 3.4 Controls of Weather and Climate Part C

C. Which of the placemarked locations in the **Altitude** folder has the highest elevation above sea level?

1. A
2. B
3. C
4. D
5. E

When you complete this part, collapse and turn off the **Altitude** folder.

Instructions for Exploration 3.4 Controls of Weather and Climate Part D:

1. Turn on and open the **Ocean Circulation** folder. The circulation of warm and cold water throughout the world's ocean basins can have a direct impact on the climate conditions along neighboring coastlines.

2. Examine each of the locations marked by placemarks *A* through *E* individually. Note their positions relative to major ocean currents.

Exploration 3.4 Controls of Weather and Climate Part D

D. Which of the locations marked in the **Ocean Circulation** folder would likely feel the greatest influence from a cold-water current?

1. A
2. B
3. C
4. D
5. E

When you complete this exploration, collapse and turn off the **3.4 – Controls of Weather and Climate** folder.

Exploration 3.4 – Controls of Weather and Climate
Short Answer

Instructions for Exploration 3.4 Controls of Weather and Climate Short Answer A:

1. Make sure you have opened the **KMZ** file from www.mygeoscienceplace.com

2. From the Places panel, expand **3. Introduction to the Atmosphere.kmz** and then open the **3.4 Controls of Weather and Climate** folder.

3. Turn on and open the **Short Answer 1** folder.

4. Double-click *Placemark 1* to fly to that location then double-click on *Placemark 2* to fly there. These placemarks are on opposite sides of the same mountain at roughly the same elevation (about 1400m). Examine the scenes at both locations and note the stark difference between the sites. Be sure to utilize appropriate zoom and navigation tools to get more detailed views of each location.

Exploration 3.4 Controls of Weather and Climate Short Answer A

A. Describe each of the locations and, based on what you've learned about the controls of weather and climate, explain why these two locations appear so different.

_____

_____

_____

_____

_____

_____

_____

Instructions for Exploration 3.4 Controls of Weather and Climate Short Answer B:

1. Use the tools in Google Earth™ to find Shanghai, China, and your current location.

Exploration 3.4 Controls of Weather and Climate Short Answer B

B. What is the elevation of both locations? What is their position relative to large bodies of water? What is their latitude? Using what you know about controls of weather and climate describe the potential differences in climate between Shanghai and your location.

_____

_____

_____

_____

_____

_____

_____

When you complete this exploration, collapse and turn off the **3.4 – Controls of Weather and Climate** folder.

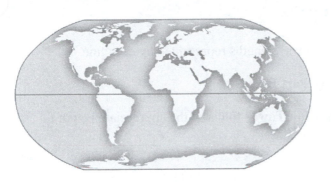

Name:_____

Date: _____

Chapter 4

Insolation and Temperature

Exploration 4.1 – Energy, Heat, and Temperature
Multiple Choice

To further our understanding of atmospheric processes, we begin by taking a closer look at the concepts of energy, heat, and temperature. Energy in the atmosphere can and does change from one form to another, and this directly impacts rates of heating and cooling. Heat can be thought of as the transfer of energy from one object to another that manifests as differences in temperature. When we use the term temperature we are referring to description of the kinetic energy of the molecules contained in a substance.

Instructions for all Parts:

1. Make sure you have opened the **KMZ** file from www.mygeoscienceplace.com

2. From the Places panel, expand **4. Insolation and Temperature.kmz** and then open the **4.1 Energy, Heat, and Temperature** folder.

Previous exercises in Google Earth™ have introduced you to a variety of imagery and data through the 3D Viewer. However, your analysis has principally utilized one portion of the electromagnetic spectrum.

Exploration 4.1 Energy, Heat, and Temperature Part A

A. When you view imagery of Earth in the Google Earth™ 3D viewer, what wavelengths of the electromagnetic spectrum are you viewing?

1. Between 500 and 1000 m
2. Between 0.4 and 0.7 μm
3. Between 1 and 100 m
4. Between 1 and 1000 μm
5. Between 0.0001 and 0.001 μm

Instructions for Exploration 4.1 Energy, Heat, and Temperature Part B:

1. Ensure that the **4.1 Energy, Heat, and Temperature** folder is open.

2. Turn on and double-click the **AM Radio Transmission Tower** folder. *Wavelength A* through *Wavelength E* illustrate potential wavelengths of radio waves emanating from the AM radio transmission tower located at this site.

3. Use the Ruler tool to measure the individual wavelengths represented by each line.

Exploration 4.1 Energy, Heat, and Temperature Part B

B. Which of the paths in the **AM Radio Transmission Tower** folder is the appropriate length for an AM radio wave?

1. Wavelength A
2. Wavelength B
3. Wavelength C
4. Wavelength D
5. Wavelength E

When you complete this part, collapse and turn off the **AM Radio Transmission Tower** folder.

Instructions for Exploration 4.1 Energy, Heat, and Temperature Part C:

1. Turn on and open the **Peak Insolation** folder. Insolation is a measure of the incoming solar radiation at any given location on Earth.

2. Fly to each of the five placemarks labeled *A* through *E* individually. Evaluate each location based on the potential insolation at each site, paying particular attention the elevation above sea level.

Exploration 4.1 Energy, Heat, and Temperature Part C

C. Which of the five placemarks in the **Peak Insolation** folder would have the highest potential insolation?

1. A
2. B
3. C
4. D
5. E

When you complete this part, collapse and turn off the **Peak Insolation** folder.

Instructions for Exploration 4.1 Energy, Heat, and Temperature Part D:

1. Turn on and open the **Temperature Scales** folder. Fahrenheit and Celsius are the two most common temperature scales for reporting temperatures around the world, though most countries favor the use of one over the other.

2. Examine the five cities labeled *A* through *E* individually. Consider the context of each city and determine what temperature scale would be most likely in each location.

Remember to turn on the *Borders and Labels* layer in the *Primary Database* to see the names of countries. Consult outside resources if necessary.

Exploration 4.1 Energy, Heat, and Temperature Part D

D. In which of the five placemarked cities in the **Temperature Scales** folder would you be more likely to find temperatures reported in degrees Fahrenheit?

1. A
2. B
3. C
4. D
5. E

When you complete this exploration, collapse and turn off the **4.1 Energy, Heat, and Temperature** folder.

Exploration 4.1 – Energy, Heat, and Temperature
Short Answer

Instructions for Exploration 4.1 Energy, Heat, and Temperature Short Answer A:

1. Turn on the *Weather* layer in the *Primary Database*.

2. View the current weather conditions for the city closest to your location. Do the same for the city of Pretoria, South Africa.

Exploration 4.1 Energy, Heat, and Temperature Short Answer A

A. Report the temperatures for both locations in Celsius, Fahrenheit, and Kelvin (you will have to calculate Kelvin). Does the temperature between these two places differ? What factors may account for that difference?

_____
_____
_____
_____
_____
_____
_____

Instructions for Exploration 4.1 Energy, Heat, and Temperature Short Answer B:

1. Turn on the *Weather* layer in the *Primary Database*.

2. Examine the weather reports for cities in the United States. Temperatures are reported in both Fahrenheit and Celsius, though we rarely use Celsius in everyday conversation. Much of the rest of the world uses the Celsius system by default and it has emerged in recent decades as the global standard, but that hasn't always been the case.

Exploration 4.1 Energy, Heat, and Temperature Short Answer B

B. When did the change to a global Celsius standard take place? What factors made the Celsius scale an advantage over other systems?

_____

_____

_____

_____

_____

_____

_____

When you complete this part, be sure to turn off the *Weather* layer in the *Primary Database*.

When you complete this exploration, collapse and turn off the **4.1 Energy, Heat, and Temperature** folder.

Exploration 4.2 – Heating and Cooling Processes
Multiple Choice

There are a number of important physical processes associated with cooling and heating of the atmosphere. In this exploration we will examine processes such as radiation, absorption, scattering, and reflection.

Instructions for all Parts:

1. Make sure you have opened the **KMZ** file from the blue box on the left.

2. From the Places panel, expand **4. Insolation and Temperature.kmz** and then open the **4.2 Heating and Cooling Processes** folder.

Instructions for Exploration 4.2 Heating and Cooling Processes Part A-B:

1. Turn on and open the **Reflection and Absorption** folder. Albedo measures the reflectivity of an object. A high albedo value means a surface reflects a high amount of incoming light.

2. Fly to the placemarks labeled *A* through *E* individually. Evaluate the albedo of each location. Do not adjust the 3D Viewer at any of the locations; rather, estimate the average albedo value for the entire landscape as shown in the view.

Exploration 4.2 Heating and Cooling Processes Part A

A. Which of the placemarked locations in the **Reflection and Absorption** folder has the highest albedo value?

1. A
2. B
3. C
4. D
5. E

Exploration 4.2 Heating and Cooling Processes Part B

B. Remaining in the **Reflection and Absorption** folder, which of the placemarked locations has the lowest albedo value?

1. A
2. B
3. C
4. D
5. E

When you complete the previous two parts, turn off and collapse the **Reflection and Absorption** folder.

Instructions Exploration 4.2 Heating and Cooling Processes Part C:

1. Double-click the hyperlink for the *St. Paul's Chapel* 360 Degree image. You should zoom into a high-resolution image of the area around St. Paul's Chapel in New York City.

2. Evaluate the scene presented in this image, paying particular attention to any evidence of transmission.

Exploration 4.2 Heating and Cooling Processes Part C

C. Which of the following statements is best supported by the view provided in the *St. Paul's Chapel* 360 Degree image?

1. The lighter-colored stone on the building that houses Staples transmits light at a higher level than the darker colored stone of St. Paul's Chapel.
2. The material that makes up the 2 Hours Free Parking sign in the Computer World window is a better transmitter than the window immediately in front of it.
3. The heat emanating from the warm asphalt street surface is the best example of transmission in this scene.
4. Of the two vehicles parked near Computer World, the silver Honda CRV has higher levels of light transmission through its automotive glass than the darker Ford Expedition's tinted windows.
5. The mirror image of the lighter-colored tall building on the darker-colored building immediately behind St. Paul's Chapel is the best example of transmission.

Instructions for Exploration 4.2 Heating and Cooling Processes Part D:

1. Double-click the hyperlink for the *Havana* 360 Degree image. You should zoom in to a high-resolution image of a street scene in Havana, Cuba.

2. Evaluate the scene presented in this image, paying particular attention to the potential for longwave radiation.

Exploration 4.2 Heating and Cooling Processes Part D

D. Which of the following items seen in the *Havana* 360 Degree image would likely emit the greatest amount of longwave radiation?

1. The bonfire.
2. The area of trees and shrubs behind the Libertad Ya sign.
3. The large bank of windows on the building with the Cuban flag.
4. The large expanse of asphalt that makes up the intersection.
5. The clouds in the sky above the scene.

When you complete this exploration, collapse and turn off the **4.2 Heating and Cooling Processes** folder.

Exploration 4.2 – Heating and Cooling Processes
Short Answer

Instructions for Exploration 4.2 Heating and Cooling Processes Short Answer A:

1. Fly to your location in Google Earth™.

2. Evaluate the surrounding landscape, taking note of any heating or cooling processes that may be present.

Exploration 4.2 Heating and Cooling Processes Short Answer A

A. Provide three examples of basic heating or cooling processes that may be present in the imagery of your location. Include a brief description of each. Be sure to note your location.

_____

_____

_____

_____

_____

_____

_____

Instructions for Exploration 4.2 Heating and Cooling Processes Short Answer B:

1. Make sure you have opened the **KMZ** file from www.mygeoscienceplace.com

2. From the Places panel, expand **4. Insolation and Temperature.kmz** and then open the **4.2 Heating and Cooling Processes** folder.

3. Open the **Gulf of Guayaquil** folder and turn on and double-click the *March 1985* image. Examine this Landsat imagery for evidence of different types of heating and cooling processes.

4. Turn on the *November 2000* image to see the same location at a different date.

5. Toggle between the two images, paying attention to any differences in heating or cooling processes present.

Exploration 4.2 Heating and Cooling Processes Short Answer B

B. Identify two locations where contrasts in the heating or cooling processes between the images are present. Identify the process and latitude and longitude where the differences can be noted.

_____

_____

_____

_____

_____

_____

_____

When you complete this exploration, collapse and turn off the **4.2 Heating and Cooling Processes** folder.

Exploration 4.3 – Variations in Heating
Multiple Choice

While a variety of physical processes work to heat or cool the atmosphere, seasonal and locational factors contribute to significant geographical variation in these processes. For example, the angle of the sun varies across the surface of the planet from one season to the next, altering the length of the day and the amount of direct heating that can occur. Properties of physical geography, such as the differing rates of heating and cooling for land and water and their relative proximity to a site, can also play a significant role in regional temperature variation.

Instructions for all Parts:

1. Make sure you have opened the **KMZ** file from www.mygeoscienceplace.com

2. From the Places panel, expand **4. Insolation and Temperature.kmz** and then open the **4.3 Variations in Heating** folder.

3. Open the **Angle of Incidence and Day Length** folder.

The angle of incidence and day length for any given location on Earth are constantly changing due to the Earth's curvature and the relative position of the Earth with the sun.

Instructions for Exploration 4.3 Variations in Heating Parts A–B:

1. Turn on the *Location 1* and *Location 2* layers.

2. Double-click each layer to fly to that site.

3. Compare the two locations and consider how the angle of incidence or day length may change at each location depending on the date.

Exploration 4.3 Variations in Heating Part A

A. In which of the following combinations of location and date would the angle of incidence be the highest?

1. Location 1, on the December Solstice
2. Location 2, on the March Equinox
3. Location 1, on the September Equinox
4. Location 2, on the December Solstice
5. Location 1, on the June Solstice

Exploration 4.3 Variations in Heating Part B

B. In which of the following combinations of location and date would the length of daylight be the longest?

1. Location 1, on the December Solstice
2. Location 2, on the March Equinox
3. Location 1, on the September Equinox
4. Location 2, on the December Solstice
5. Location 1, on the June Solstice

When you complete the previous two parts, turn off and collapse the **Angle of Incidence and Day Length** folder.

Instructions for Exploration 4.3 Variations in Heating Part C:

1. Double-click the *Scotts Bluff, Nebraska*, tour to see a 360-degree view of this famous landmark in western Nebraska. Examine the surrounding landscape using Google Earth™.

2. Turn off the *Scotts Bluff, Nebraska*, tour controls.

3. Turn on and double-click the *Scotts Bluff 1* placemark. The Sun tool will automatically turn on, giving you a view of this location illuminated at a particular date and time of day.

4. Turn off the *Scotts Bluff 1* placemark then turn on and double-click *Scotts Bluff 2* placemark and repeat the same examination. Turn off *Scotts Bluff 2* then turn on and double-click *Scotts Bluff 3* to examine the final scene.

5. Analyze the shadows that are cast by Scotts Bluff in each of the three placemarks, making sure that only one placemark is turned on at a time. Be sure to use only the shadows cast by the profile of Scotts Bluff in your assessment.

Exploration 4.3 Variations in Heating Part C

C. Which of the following statements is best supported by the evidence seen in the three numbered Scotts Bluff placemarks?

1. *Scotts Bluff 1* represents conditions during the middle of the day in the summer.
2. *Scotts Bluff 2* represents conditions during the early morning of a winter day.
3. *Scotts Bluff 3* represents conditions during the late afternoon.
4. *Scotts Bluff 2* represents conditions at mid-day on the summer solstice.
5. *Scotts Bluff 1* represents conditions at mid-day on the winter solstice.

IMPORTANT: When you complete this part, turn off the Sun tool.

Instructions for Exploration 4.3 Variations in Heating Part D:

1. Turn on and double-click the *US Solar Potential* layer. The *US Solar Potential* layer shows the levels of photovoltaic solar radiation. Angle of incidence and day length are two of the major factors, along with average cloudiness, contributing to the level of insolation and potential for solar energy development.

2. Evaluate the Photovoltaic Solar Radiation potential of the United States, paying particular attention to the regions with the highest potential for solar development.

Exploration 4.3 Variations in Heating Part D

D. Which of the following cities has the highest level of photovoltaic solar radiation based on data in the *US Solar Potential* layer?

1. Orlando, Florida
2. San Diego, California
3. El Paso, Texas
4. Minneapolis, Minnesota
5. Cleveland, Ohio

When you complete this exploration, collapse and turn off the **4.3 Variations in Heating** folder.

Exploration 4.3 – Variations in Heating
Short Answer

Instructions for all Parts:

1. Make sure you have opened the **KMZ** file from www.mygeoscienceplace.com

2. From the Places panel, expand **4. Insolation and Temperature.kmz** and then open the **4.3 Variations in Heating** folder.

3. Open the **Angle of Incidence and Day Length** folder.

Instructions for Exploration 4.3 Variations in Heating Short Answer A:

1. Turn on and double-click the *US Solar Potential* layer.

2. Evaluate the levels of photovoltaic solar radiation at your location.

Exploration 4.3 Variations in Heating Short Answer A

A. Identify you location. Do you live in an area of lower, moderate, or higher potential for development of this resource? What are the local characteristics (e.g., climate, latitude, weather) that may contribute to your location's rating?

_____

_____

_____

_____

_____

_____

_____

When you complete this part, turn off the *US Solar Potential* layer.

Instructions for Exploration 4.3 Variations in Heating Short Answer B:

1. Turn on and double-click on the *Day Length* placemark. The time slider located in the upper left corner of the 3D Viewer allows you to change the date and time.

2. Move the time slider using the forward and reverse icons to watch the progression of daylight and darkness across the planet. Pay particular attention to the angle of the sun on the surface of Earth and the distribution of light (day) and dark (night) areas.

Exploration 4.3 Variations in Heating Short Answer B

B. Explain what you see relative to the date shown in the time slider (12/20/2011) and the status of sun angle and day length at that time of year.

_____

_____

_____

_____

_____

_____

_____

When you complete this part, turn off the Sun tool.

When you complete this evaluation, collapse and turn off the **4.3 Variations in Heating** folder.

Exploration 4.4 – Heat Transfer, Vertical Temperature Patterns, and Global Warming Multiple Choice

In this exploration, we examine how heat transfer moves surplus heat pole-ward from the equator by circulation patterns in the atmosphere in oceans. Though this heat transfer often concentrates on horizontal transfers, there is also a vertical temperature gradient; an examination of environmental lapse rates reveals these generalized patterns. We conclude this exploration with a brief look at factors that enhance the Earth's natural greenhouse effect, resulting in a process commonly referred to as global warming.

Instructions for all Parts:

1. Make sure you have opened the **KMZ** file from www.mygeoscienceplace.com

2. From the Places panel, expand **4. Insolation and Temperature.kmz** and then open the **4.4 Heat Transfer, Vertical Temperature Patterns, and Global Warming** folder.

Instructions for Exploration 4.4 Heat Transfer, Vertical Temperature Patterns, and Global Warming Part A:

1. Expand the *Sea Surface Temperature* layer then turn on the *SST* sub-layer.

2. Click the hyperlinked name of the *Sea Surface Temperature* layer to learn more about the layer that shows average sea surface temperature readings around the globe. Oceanic circulation is a major component of global heat transfer. Large-scale oceanic water movements, or ocean currents, are a result of air blowing over the surface of the water. Coastal areas are profoundly influenced by the presence of cool or warm ocean currents. Cities at similar latitudes may have different climates primarily as a result of the temperature of nearby ocean currents.

3. Examine the pattern of sea surface temperature, paying particular attention to the presence of warm or cool water near coastal areas.

Exploration 4.4 Heat Transfer, Vertical Temperature Patterns, and Global Warming Part A

A. Based on the data in the *Sea Surface Temperature* layer, which of the cities listed below would have the warmest sea surface temperatures nearby? If you are not familiar with the location of these cities, utilize the Search panel.

1. Wellington, New Zealand
2. Perth, Australia
3. Lisbon, Portugal
4. Georgetown, Guyana
5. Halifax, Canada

When you complete this part, turn off and collapse the *Sea Surface Temperature* layer.

Instructions for Exploration 4.4 Heat Transfer, Vertical Temperature Patterns, and Global Warming Part B:

1. Turn on and double-click the **Average Lapse Rate** folder. The average lapse rate is the global average temperature change experienced with an increase or decrease in altitude, using a rate of 6.5 °C per 1000 meters for any location within the troposphere. The Mount Evans Scenic Byway winds from the town of Idaho Springs, Colorado, to the top of nearby Mount Evans, one of the state's "14ers" – or mountains over 14,000 feet in elevation.

2. Examine the five placemarks located along the Mount Evans Scenic Byway and labeled *A* through *E*. Note the elevation of each placemark and its relative change compared to the starting elevation of the route at 2300 meters in Idaho Springs.

Exploration 4.4 Heat Transfer, Vertical Temperature Patterns, and Global Warming Part B

B. At which of the placemarked locations would you encounter a temperature that is approximately 10 °C colder than the temperature experienced in Idaho Springs?

1. A
2. B
3. C
4. D
5. E

When you complete this part, turn off and collapse the **Average Lapse Rate** folder.

Instructions for Exploration 4.4 Heat Transfer, Vertical Temperature Patterns, and Global Warming Part C:

1. Turn on and double-click the **Global Warming** folder. This displays data from the United Kingdom's Meteorology Office. This particular imagery models the potential global temperature variation with the premise of a 4 °C average rise in global temperatures.
2. Examine the imagery noting that some areas of the globe will likely warm much more than others.

Exploration 4.4 Heat Transfer, Vertical Temperature Patterns, and Global Warming Part C

C. Which of the following global regions exhibits the greatest potential increase in temperature, based on the model imagery in the **Global Warming** folder?

1. Antarctica
2. Arctic
3. Europe
4. Sub-Saharan Africa
5. South Asia

When you complete this part, turn off the **Global Warming** folder.

Instructions for Exploration 4.4 Heat Transfer, Vertical Temperature Patterns, and Global Warming Part D:

1. Turn on and open the **Greenhouse Gasses** folder. Global temperature increases can occur when the concentration of specific greenhouse gasses in the atmosphere rise. Gas emissions may be the result of naturally occurring events or anthropogenic activity.

2. Examine the placemarks labeled *A* through *E*. Note the potential for each location to either increase or decrease the release of atmospheric greenhouse gasses.

Exploration 4.4 Heat Transfer, Vertical Temperature Patterns, and Global Warming Part D

D. Which of the placemarked locations would contribute the **least** amount of greenhouse gasses?

1. A
2. B
3. C
4. D
5. E

When you complete this exploration, turn off and collapse the **4.4 Heat Transfer, Vertical Temperature Patterns, and Global Warming** folder.

**Exploration 4.4 – Heat Transfer, Vertical Temperature Patterns, and Global Warming
Short Answer**

Instructions for all Parts:

1. Make sure you have opened the **KMZ** file from www.mygeoscienceplace.com

2. From the Places panel, expand **4. Insolation and Temperature.kmz** and then open the
   4.4 Heat Transfer, Vertical Temperature Patterns, and Global Warming folder.

**Instructions for Exploration 4.4 Heat Transfer, Vertical Temperature Patterns, and
Global Warming Short Answer A:**

1. Click the hyperlink under *Science on a Sphere* to open a website for the National Oceanic
   and Atmospheric Administration's digital Earth science education program. This page
   lists available KML files to be used in Google Earth™.

2. Click the *Ocean Currents* file to begin the download. Once the file has downloaded to
   your computer, click it to open the data in Google Earth™. A layer called *Ocean
   Currents* will load in the *Temporary Places* section of the Places panel.

3. Click the *Ocean Currents* layer title to open an information panel from NOAA's Science
   On a Sphere® and learn more about this layer, which shows the relative speed of moving
   ocean water. The dark blues indicate the slowest moving water, the red indicates the
   fastest water, while the colors in between indicate a range of progressively changing
   water speeds.

4. Compare this map of ocean current speed to the map of major surface ocean currents in
   your text.

**Exploration 4.4 Heat Transfer, Vertical Temperature Patterns, and Global Warming
Short Answer A**

A. What correlation can you make between the speed of ocean currents and the temperature of the
water in those currents? In your response, be sure to indicate at least two currents that support
your response.

_____

_____

_____

_____

_____

_____

_____

When you complete this part, turn off the *Ocean Currents* layer.

Instructions for Exploration 4.4 Heat Transfer, Vertical Temperature Patterns, and Global Warming Short Answer B

1. Open the *Sea Surface Temperature* layer.

2. Turn on the *SST Anomaly* sub-layer. This data shows the difference between average sea surface temperature readings in any given month and the average sea surface temperature for all of those months in prior years. Click the *Sea Surface Temperature* layer link to open a pop-up with additional information.

3. Evaluate the *SST Anomaly* layer paying particular attention to regions with large difference in either the positive or negative direction. Use the Historical Imagery tool to show the most recent data.

Exploration 4.4 Heat Transfer, Vertical Temperature Patterns, and Global Warming Short Answer B

B. What coastal locations have the most significant sea surface temperature anomalies above and below normal based on the most current data available?

_____
_____
_____
_____
_____
_____
_____

When you complete this exploration, turn off and collapse the **4.4 Heat Transfer, Vertical Temperature Patterns, and Global Warming** folder.

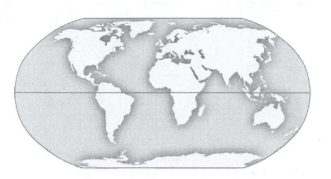

Encounter Physical Geography

Name:_____

Date: _____

Chapter 5
Atmospheric Pressure and Wind

Exploration 5.1 – Pressure
Multiple Choice

The most direct reflection of atmospheric pressure is the horizontal movement of air known as wind. Significant differences in areas of high and low pressure can lead to high winds. Oftentimes the most significant winds are associated with transient atmospheric features such as low-pressure systems and associated storms.

Instructions for all Parts:

1. Make sure you have opened the **KMZ** file from www.mygeoscienceplace.com

2. From the Places panel, expand **5. Atmospheric Pressure and Wind.kmz** and then open the **5.1 Pressure** folder.

Instructions for Exploration 5.1 Pressure Parts A–C:

1. Turn on and double-click the **Isobar** folder. The map included in this folder shows the atmospheric pressure conditions at 12 UTC on February, 2, 2011.

2. Examine the location of the five placemarks labeled *A* through *E* on the map, paying attention to the relative position of high-pressure and low-pressure zones as well as any significant gradients between systems.

Exploration 5.1 Pressure Part A

A. Which of the following placemarks would have the lowest atmospheric pressure reading?

   1. A
   2. B
   3. C
   4. D
   5. E

Exploration 5.1 Pressure Part B

B. Which of the following placemarks would likely have the highest wind speeds?

1. A
2. B
3. C
4. D
5. E

Exploration 5.1 Pressure Part C

C. At which of the following placemarks would surface winds be most likely to come from the east?

1. A
2. B
3. C
4. D
5. E

When you complete the previous three questions, collapse and turn off the **Isoline** folder.

Instructions for Exploration 5.1 Pressure Part D:

1. Turn on and double-click the *Armistice Day Blizzard of 1940* layer.

2. Examine the features of a strong blizzard that affected large parts of the country in November 1940. Pay particular attention to the patterns of high and low pressure and the gradients between systems.

Exploration 5.1 Pressure Part D

D. Which of the following statements is most strongly supported by the information presented in the *Armistice Day Blizzard of 1940* layer?

1. Atmospheric pressure in the core of the low-pressure area was less than 972 mb.
2. The pressure gradient across the northern Great Plains exceeded 100 mb.
3. The highest pressure gradient associated with the storm at the time of this map was located in Minnesota.
4. A ridge of high pressure roughly follows the Mississippi River valley.
5. The area of highest pressure on the map is located in central Texas.

When you complete this exploration, collapse and turn off the **5.1 Pressure** folder.

Exploration 5.1 – Pressure
Short Answer

Instructions for Exploration 5.1 Pressure Short Answer A:

1. Make sure you have opened the **KMZ** file from www.mygeoscienceplace.com

2. From the Places panel, expand **5. Atmospheric Pressure and Wind.kmz** and then open the **5.1 Pressure** folder.

3. Turn on and double-click the *Armistice Day Blizzard of 1940* layer.

4. Double-click the hyperlink associated with this layer to open a National Weather Service summary of this storm.

5. Read about the wide-ranging impacts of the Armistice Day Blizzard of 1940. As you read this summary, think about an event that occurred at your location and consider the atmospheric conditions that might have been present to create it.

Exploration 5.1 Pressure Short Answer A

A. Describe a significant weather event that occurred in your area. Tell when it occurred and what some of the most significant impacts were. In your response, be sure to include a hypothesis concerning the role of atmospheric pressure and pressure gradients in the scale and severity of the event. You might consider checking the website for your local office of the National Weather Service. Be sure to note your location in your response.

_____

_____

_____

_____

_____

_____

_____

When you complete this part, turn off the *Armistice Day Blizzard of 1940* layer.

Instructions for Exploration 5.1 Pressure Short Answer B:

1. Turn on the *Weather* layer in the *Primary Database*.

Exploration 5.1 Pressure Short Answer B

B. Based on the cloud cover and precipitation, make a hypothesis about the location of cyclonic and anti-cyclonic circulations in North America and Europe. Be sure to report the day and time of your observations.

_____

_____

_____

_____

_____

_____

_____

When you complete this exploration, collapse and turn off the **5.1 Pressure** folder.

5.2 – General Circulation
Multiple Choice

Earth's atmosphere is an extraordinarily complex feature that is in constant movement. Movement in the atmosphere ranges from the molecular level to local, regional, and national scales. This exploration examines the general patterns of circulation that occur around the globe.

Instructions for all Parts:

1. Make sure you have opened the **KMZ** file from www.mygeoscienceplace.com

2. From the Places panel, expand **5. Atmospheric Pressure and Wind.kmz** and then open the **5.2 General Circulation** folder.

Instructions for Exploration 5.2 General Circulation Part A:

1. Turn on and double-click the **Fukushima** folder. This image shows global precipitation for the days immediately following the March 2011 Japan earthquake, tsunami, and nuclear reactor accident.

2. Examine the information presented in this map, paying attention to the patterns of general atmospheric circulation as evidenced by precipitation relative to the location of the Fukushima Nuclear Power Plant.

Exploration 5.2 General Circulation Part A

A. Based on your knowledge of the general circulation of the atmosphere and the data in the *Fukushima* layer, which of the following coordinate locations was in the highest risk area for downwind radiation contamination in the days after the March 2011 event?

1. 27°N, 118°E
2. 3°N, 123°E
3. 37°N, 153°E
4. 65°N, 112°E
5. 37°N, 127°E

When you complete this part, collapse and turn off the **Fukushima** folder.

Instructions for Exploration 5.2 General Circulation Part B:

1. Turn on and double-click the *Dust Plumes* layer. This image shows a section of the southwest African coast. The dust plumes are created by winds blowing across the Namib Desert.

2. Examine the patterns seen in the dust plumes, paying particular attention to their orientation.

Exploration 5.2 General Circulation Part B

B. What component of global circulation is most responsible for the prevailing winds in this region?

1. Santa Ana Winds
2. Northern Hemisphere Westerlies
3. Subtropical High
4. Southeast Trades
5. Sea Breeze

When you complete this part, turn off the *Dust Plumes* layer.

Instructions for Exploration 5.2 General Circulation Part C:

1. Open and turn on the **Lost at Sea** folder.

2. Examine each of the five placemarks labeled *A* through *E*, noting their location relative to general global circulation patterns. Consider a scenario in which you were lost at sea on a raft with only a small sail for propulsion and had to rely on the wind at each of these locations.

Exploration 5.2 General Circulation Part C

C. At which of the placemarked locations would you have the **least** chance of finding substantial wind during the month of March?

1. A
2. B
3. C
4. D
5. E

When you complete this part, collapse and turn off the **Lost at Sea** folder.

Instructions for Exploration 5.2 General Circulation Part D:

1. Use the Search Panel of Google Earth™ to locate the following cities: Khartoum, Sudan; Sana'a, Yemen; N'Djamena, Chad; Lusaka, Zambia; and Asmera, Eritrea.

2. Note the location of each city, making sure to consider the seasonal variations in wind and circulation that may be present at each.

Exploration 5.2 General Circulation Part D

D. Which of the following cities would most likely experience the effects of the ITCZ in December?

1. Khartoum, Sudan
2. Sana'a, Yemen
3. N'Djamena, Chad
4. Lusaka, Zambia
5. Asmera, Eritrea

When you complete this exploration, collapse and turn off the **5.2 General Circulation** folder.

Exploration 5.2 – General Circulation
Short Answer

Instructions for Exploration 5.2 General Circulation Short Answer A:

1. Turn on the *Weather* layer in the *Primary Database*.

2. Examine the cloud and circulation patterns found throughout the ITCZ.

Exploration 5.2 General Circulation Short Answer A

A. What regions are currently seeing impacts from the ITCZ? Note the date you are examining this information and provide the name and location of three cities that currently fall within that zone. What would the weather be like in these cities?

_____

_____

_____

_____

_____

_____

_____

When you complete this part, turn off the *Weather* layer.

Exploration 5.2 General Circulation Short Answer B

B. In what general global circulation band do you live? What tools or functions in Google Earth™ could you utilize to verify that this is the case?

_____

_____

_____

_____

_____

_____

_____

Exploration 5.3 – Local Wind Systems
Multiple Choice

Unique local circumstances related to topography and proximity to the ocean can contribute to the creation of localized wind systems. These lesser winds can contribute to hazards such as fire or fog but can also be utilized as an energy resource.

Instructions for all Parts:

1. Make sure you have opened the **KMZ** file from www.mygeoscienceplace.com

2. From the Places panel, expand **5. Atmospheric Pressure and Wind.kmz** and then open the **5.3 Local Wind Systems** folder.

Instructions for Exploration 5.3 Local Wind Systems Part A:

1. Turn on and double-click the *San Francisco* layer. The vantage point seen here looks east through the Golden Gate toward the city of San Francisco.

2. Click the first hyperlink to view the YouTube video titled "Fog at Golden Gate Bridge Time Lapse." This video shows fog rolling through the Golden Gate from the same perspective seen in the Google Earth™ 3D Viewer.

3. The second YouTube video shows a similar scene from a location closer to the bridge.

4. Examine the movement of fog seen in the two videos, making sure to note the relative movement of fog.

Exploration 5.3 Local Wind Systems Part A

A. What localized wind pattern is seen in the videos associated with the *San Francisco* layer?

1. Land breeze
2. Sea breeze
3. Chinook
4. Foehn wind
5. Valley breeze

When you complete this part, turn off the *San Francisco* layer.

Instructions for Exploration 5.3 Local Wind Systems Part B:

1. Turn on and double-click the *National Renewable Energy Laboratory* layer.

2. Click the hyperlink to view a map showing the 50-meter wind power resource potential in Colorado.

3. Examine the map, making sure to note the areas where wind resource potential is the highest.

Exploration 5.3 Local Wind Systems Part B

B. Based on the data in the NREL 50 m Wind Power Resource map, what locations have outstanding or superb wind resource potential?

1. The windward slopes of the Rocky Mountains.
2. Along major river valleys.
3. Near the largest cities.
4. The leeward slopes of the Rocky Mountains.
5. The western third of the state.

When you complete this part, turn off the *National Renewable Energy Laboratory* layer.

Instructions for Exploration 5.3 Local Wind Systems Part C:

1. Open the **Wind Energy** folder.

2. Each of the placemarks labeled *A* through *E* represents locations of existing wind energy developments. Double-click each of the sites individually and examine the landscape these wind turbines occupy. Be sure to pay attention to the potential local wind patterns that contribute to the rotation of the turbines.

Exploration 5.3 Local Wind Systems Part C

C. At which of the five placemarked locations would wind turbines be driven by a land breeze/sea breeze dynamic?

1. A
2. B
3. C
4. D
5. E

Instructions for Exploration 5.3 Local Wind Systems Part D:

1. Turn on and double-click the **Santa Ana Winds** folder. The *Santa Ana Fires* photo is a MODIS satellite image that shows plumes of smoke coming from a series of fires in southern California. Fires here are often exacerbated by local Santa Ana winds.

2. Turn on each of the labeled placemarks *A* through *E*. Note the position of each relative to the location of the fires and smoke plumes.

3. Examine the evidence of regional air circulation seen in the image and the relative location of each of the placemarks.

Exploration 5.3 Local Wind Systems Part D

D. Where is the most likely location of a cell of high pressure in this scenario?

1. A
2. B
3. C
4. D
5. E

When you complete this exploration, collapse and turn off the **5.3 Local wind Systems** folder.

Exploration 5.3 – Local Wind Systems
Short Answer

Instructions for all Parts:

1. Make sure you have opened the **KMZ** file from www.mygeoscienceplace.com

2. From the Places panel, expand **5. Atmospheric Pressure and Wind.kmz** and then open the **5.3 Local Wind Systems** folder.

Instructions for Exploration 5.3 Local Wind Systems Short Answer A:

1. Turn on and double-click the *Central US Fires* image. This MODIS satellite image shows a series of fires in the Great Plains of the United States stretching from southeast Nebraska into eastern Texas.

2. Examine the evidence of circulation patterns seen in the image, paying particular attention to the smoke plumes generated by the fires. Be certain to keep in mind the typical circulation patterns that exist in this part of North America.

Exploration 5.3 Local Wind Systems Short Answer A

A. Based on the smoke plumes seen in this image, do you see any evidence of local wind patterns, or are regional circulation patterns dominant? Form a hypothesis for the location of surface high- and low-pressure systems. Where might these be located? What clues led you to your hypothesis?

_____
_____
_____
_____
_____
_____
_____

Instructions for Exploration 5.3 Local Wind Systems Short Answer B:

1. Double-click the *Beach* placemark to fly into a Street Scene view in Fort Lauderdale, Florida.

Exploration 5.3 Local Wind Systems Short Answer B

B. What type of local wind pattern would be likely in this location at this time? What evidence leads you to that conclusion?

_____
_____
_____
_____
_____
_____
_____

When you complete this exploration, collapse and turn off the **5.3 Local Wind Systems** folder.

Exploration 5.4 – El Niño
Multiple Choice

El Niño is a complex phenomenon that illustrates the interrelated nature of the ocean and the atmosphere. In this cyclical event, surface waters off the west coast of South America become abnormally warm. This creates a series of impacts and effects that are felt in a number of locations around the globe.

Instructions for all Parts:

1. Make sure you have opened the **KMZ** file from www.mygeoscienceplace.com

2. From the Places panel, expand **5. Atmospheric Pressure and Wind.kmz** and then open the **5.4 El Niño** folder.

Instructions for Exploration 5.4 El Niño Part A:

1. Open the **El Niño 1997** folder.

2. Turn on and double-click the **Sea Surface Temperature Anomalies** sub-folder. This folder contains information showing the departure from normal for sea surface temperatures in the equatorial Pacific.

3. Use the historical imagery animation to show the change over time for sea surface temperatures from 1997 through 1999. (Note: Depending on your connection speed, the images for each available date may take some time to load.)

Exploration 5.4 El Niño Part A

A. Which of the following months had the highest sea surface anomaly from 1997 through 1999?

1. January 1997
2. May 1997
3. December 1997
4. August 1998
5. December 1998

Instructions for Exploration 5.4 El Niño Part B:

1. Ensure that the **El Niño 1997** folder is open and all sub-folders are turned off.

2. Turn on and double-click the **Sea Surface Height Anomaly** sub-folder. This image shows the departure from normal for sea surface heights in the equatorial Pacific. Changes in sea surface heights are related to the expansion and contraction of water as it changes temperature.

3. Use the historical imagery animation to show the change over time for sea surface heights from 1997 through 1999. (Note: Depending on your connection speed, the images for each available date may take some time to load.)

Exploration 5.4 El Niño Part B

B. Which of the following statements is best supported from the evidence seen in the **Sea Surface Height Anomaly** sub-folder data?

1. Abnormally high sea surface heights are generally correlated with abnormally high sea surface temperatures.
2. Abnormally high sea surface heights are generally correlated with abnormally low sea surface temperatures.
3. During the height of an El Niño event, we would expect to see sea surface heights off the coast of South America 20 to 30 cm below their typical heights.
4. During the height of an El Niño event, we would expect to see sea surface heights in the western Pacific 20 to 30 cm above their typical heights.
5. No relationship between sea surface height and sea surface temperature is discernible.

When you complete this part, turn off and collapse the **El Niño 1997** folder.

Instructions for Exploration 5.4 El Niño Part C:

1. Turn on and double-click the *Unusually Intense Monsoon Rains* layer.

2. This NASA-generated layer shows the rainfall anomaly between the first week of August 2010 compared with normal precipitation rates. In early August 2010, an exceptionally strong monsoon event began in parts of south Asia.

3. Examine the rainfall anomaly data, paying particular attention to the patterns seen over regions of south Asia.

Exploration 5.4 El Niño Part C

C. Which of the following physical regions of south Asia had the greatest increase in precipitation from normal during the August 2010 monsoon?

1. Ganges Delta
2. Himalaya Mountains
3. Hindu Kush
4. Indus Delta
5. Western Ghats

When you complete this part, turn off and collapse the *Unusually Intense Monsoon Rain* layer.

Instructions for Exploration 5.4 El Niño Part D:

1. Turn on and open the **Global Consequences** folder.

2. Click the hyperlink to view a discussion of the global consequences of El Niño during winter and summer months.

3. Evaluate the patterns in the data paying particular attention to the impacts seen at the five areas placemarked in the **Global Consequences** folder.

Exploration 5.4 El Niño Part D

D. Which of the regions labeled in the **Global Consequences** folder would experience the most dry conditions during the summer of El Niño summers?

1. The Maghreb
2. Iberian Peninsula
3. Arabian Peninsula
4. Horn of Africa
5. Isthmus of Panama

When you complete this exploration, collapse and turn off the **5.4 El Niño** folder.

Exploration 5.4 – El Niño
Short Answer

Instructions for all Parts:

1. Make sure you have opened the **KMZ** file from www.mygeoscienceplace.com

2. From the Places panel, expand **5. Atmospheric Pressure and Wind.kmz** and then open the **5.4 El Niño** folder.

Instructions for Exploration 5.4 El Niño Short Answer A:

1. Turn on and double-click the *Unusually Intense Monsoon Rains* layer.

2. This layer shows the impacts of monsoon rains in south Asia during a La Niña year. La Niña conditions typically produce extreme monsoon events, and the 2010 season was no different. While El Niño/La Niña patterns can have an impact on the severity of the monsoon, this process is a regular event in south Asia as well as in other parts of the world.

Exploration 5.4 El Niño Short Answer A

A. Explain the basic process that occurs during a monsoon event in south Asia. Be sure to note the location of high- and low-pressure zones in the Indian Ocean and over central Asia during the spring and fall.

_____

_____

_____

_____

_____

_____

_____

When you complete this part, turn off the *Unusually Intense Monsoon Rains* layer.

Instructions for Exploration 5.4 El Niño Short Answer B:

1. Turn on and double-click the *NOAA Climate Prediction Center* placemark.

2. Click the hyperlink below the layer name to find a description of impacts of El Niño across the United States.

3. Under the Rank Maps section, click the December-February links for both Temperature and Precipitation individually. Examine each, being sure to note the conditions at your location and the extreme locations across the nation.

Exploration 5.4 El Niño Short Answer B

B. Based on the evidence from the Rank Maps of El Niño Impacts on the United States, Is your location in an extreme area? Where do the extremes occur?

_____

_____

_____

_____

_____

_____

_____

When you complete this exploration, collapse and turn off the **5.4 El Niño** folder.

Encounter Physical Geography

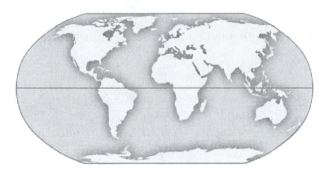

Name:_____

Date: _____

Chapter 6
Atmospheric Moisture

Exploration 6.1 – Atmospheric Moisture
Multiple Choice

This exploration examines water vapor, the source of moisture for clouds and precipitation in the atmosphere. The distribution of water vapor in Earth's atmosphere can vary substantially by location on the globe and altitude above the surface. Water vapor in the atmosphere is also affected by rates of evaporation, sublimation, and condensation.

Instructions for all Parts:

1. Make sure you have opened the **KMZ** file from www.mygeoscienceplace.com

2. From the Places panel, expand **6. Atmospheric Moisture.kmz** and then open the **6.1 Atmospheric Moisture** folder.

Instructions for Exploration 6.1 Atmospheric Moisture Part A:

1. Open and turn on the **Evaporation** folder.

2. Examine each of the five placemarks labeled *A* through *E*, noting the potential for evaporation at each location. Be sure to think about potential evaporation factors including high temperatures, low relative humidity, and high winds.

Exploration 6.1 Atmospheric Moisture Part A

A. Which of the five locations placemarked in the **Evaporation** folder would have the highest potential evaporation?

1. A
2. B
3. C
4. D
5. E

When you complete this part, collapse and turn off the **Evaporation** folder.

Instructions for Exploration 6.1 Atmospheric Moisture Part B:

1. Turn on and open the **Sublimation** folder.

2. Examine the five placemarks labeled *A* through *E*, noting the potential for sublimation of water at each location.

Exploration 6.1 Atmospheric Moisture Part B

B. Which location labeled in the **Sublimation** folder would have the highest possibility for sublimation of water?

1. A
2. B
3. C
4. D
5. E

When you complete this part, turn off the **Sublimation** folder.

Instructions for Exploration 6.1 Atmospheric Moisture Part C–D:

1. Turn on and open the **Water Vapor** folder. The two layers in this folder show the atmospheric water vapor during the months of January and June, in centimeters of precipitable water.

2. Examine the shifting patterns of atmospheric water vapor between the two seasonal datasets by turning on and off the two data layers. Pay particular attention to areas with high and low water vapor values and regions where water vapor varies significantly between the two seasons. Use the Search panel to locate places with which you are unfamiliar.

Exploration 6.1 Atmospheric Moisture Part C

C. Which of the following physical regions has the highest amount of atmospheric water vapor during the month of January based on the data in the **Water Vapor** folder?

1. Rocky Mountains
2. Cape York Peninsula
3. Arabian Peninsula
4. Antarctica
5. Andes Mountains

Exploration 6.1 Atmospheric Moisture Part D

D. Which of the following statements is best supported by the data in the two data layers in the **Water Vapor** folder?

1. The Amazon Basin exhibits significant fluctuations in water vapor between the two datasets.
2. The Tibetan Plateau exhibits significant fluctuations in water vapor between the two datasets.
3. Indonesia exhibits significant fluctuations in water vapor between the two datasets.
4. The lower Indus River valley exhibits significant fluctuations in water vapor between the two datasets.
5. Greenland exhibits significant fluctuations in water vapor between the two datasets.

When you complete this exploration, collapse and turn off the **6.1 Atmospheric Moisture** folder.

Exploration 6.1 – Atmospheric Moisture
Short Answer

Instructions for all Parts:

1. Make sure you have opened the **KMZ** file from www.mygeoscienceplace.com

2. From the Places panel, expand **6. Atmospheric Moisture.kmz** and then open the **6.1 Atmospheric Moisture** folder.

Instructions for Exploration 6.1 Atmospheric Moisture Short Answer A:

1. Turn on and open the **Water Vapor** folder.

2. Examine seasonal changes in the areas around the lower Indus River Valley and west Africa south of the Sahara Desert by turning the January and June layers on and off.

Exploration 6.1 Atmospheric Moisture Short Answer A

A. Explain the phenomena that accounts for each of these regions having a dramatic seasonal change in atmospheric water vapor. Identify one other location impacted by a similar phenomenon that also displays a significant change in seasonal water vapor values.

_____

_____

_____

_____

_____

_____

_____

When you complete this part, collapse and turn off the **Water Vapor** folder.

Instructions for Exploration 6.1 Atmospheric Moisture Short Answer B:

1. Turn on and double-click the **Relative Humidity** folder. The current relative humidity values for North America are displayed in a color-coded scheme.

2. Open the *Weather* layer in the *Primary Database*, and then turn on the *Clouds* and *Radar* layers.

Exploration 6.1 Atmospheric Moisture Short Answer B

B. Explain the relationship between relative humidity values and current cloud and precipitation values in North America.

_____

_____

_____

_____

_____

_____

_____

When you complete this exploration, collapse and turn off the **6.1 Atmospheric Moisture** folder.

Exploration 6.2 – Precipitation
Multiple Choice

Rain, snow, and other forms of precipitation are commonplace in the lower portion of the atmosphere. However, the distribution of precipitation in its varied forms is uneven. In this exploration, we utilize Google Earth™ to highlight the patterns of precipitation.

Instructions for all Parts:

1. Make sure you have opened the **KMZ** file from www.mygeoscienceplace.com.

2. From the Places panel, expand **6. Atmospheric Moisture.kmz** and then open the **6.2 Precipitation** folder.

Instructions for Exploration 6.2 Precipitation Parts A–B:

1. Turn on and double-click the **Annual US Precipitation** folder.

2. Examine the pattern of precipitation across the United States, paying particular attention to regional patterns and areas with dramatic shifts in precipitation values.

Exploration 6.2 Precipitation Part A

A. Which of the following statements is most strongly supported by the data presented in the **Annual US Precipitation** folder?

1. All precipitation values in the western United States are lower than the eastern United States.
2. The most dramatic changes in precipitation over local regions generally occur around mountain ranges.
3. In general, precipitation values in the Great Plains decline from west to east.
4. States with the highest precipitation variability are generally located in the southeastern United States.
5. The lowest precipitation values in the United States are found in the White Sands Desert of New Mexico.

Exploration 6.2 Precipitation Part B

B. Which of the following locations shows evidence of a rain shadow?

1. North of the Chihuahuan Desert
2. East of the Ouachita Mountains
3. East of the Cascade Range
4. East of Lake Michigan
5. East of the Blue Ridge Mountains

When you complete these parts, turn off the **Annual US Precipitation** folder.

Instructions for Exploration 6.2 Precipitation Part C:

1. Locate the *January 2011 Precipitation* and *June 2011 Precipitation* layers. These layers display the global pattern of precipitation during January and June 2011.

2. Zoom in to South America and toggle between the January and June layers to examine the changes that occurred between these two seasons across the continent.

Exploration 6.2 Precipitation Part C

C. Which of the following statements is best supported by the data in the *January 2011 Precipitation* and *June 2011 Precipitation* layers?

1. The greatest variation is found in Colombia.
2. Areas with lowest levels of precipitation in both January and June are found in Brazil.
3. The region with the highest levels of precipitation in both January and June are found in Paraguay.
4. Greater precipitation falls continent-wide in January rather than June.
5. January has greater variation in precipitation across the continent than June.

When you complete this part, turn off both the *January* and *June 2011 Precipitation* layers.

Instructions for Exploration 6.2 Precipitation Part D:

1. Turn on and double-click the **Snow Depth** folder. This data displays the recorded snow depth in late January 2012.

2. Evaluate the pattern of snow depth across the United States and Canada, paying particular attention to the areas of highest and lowest snow depth and correlation between physical terrain.

Exploration 6.2 Precipitation Part D

D. Which of the following statements is best supported by the data presented in the **Snow Depth** folder?

1. The highest snowfall totals are found in the central Rocky Mountains.
2. Stream and river valleys are evident in areas with snow by their higher snow depth values.
3. No areas of the Great Plains have a snow depth greater than 10 cm.
4. No measurable snow on the ground south of Washington, DC.
5. Snow depths are higher on the eastern shore of Lake Michigan than on the western shore.

When you complete this exploration, collapse and turn off the **6.2 Precipitation** folder.

Exploration 6.2 – Precipitation
Short Answers

Instructions for all Parts:

1. Make sure you have opened the **KMZ** file from www.mygeoscienceplace.com

2. From the Places panel, expand **6. Atmospheric Moisture**.kmz and then open the **6.2 Precipitation** folder.

Instructions for Exploration 6.2 Precipitation Short Answer A:

1. Turn on the **Annual US Precipitation** folder. This layer shows the annual precipitation across the United States.

Exploration 6.2 Precipitation Short Answer A

A. Based on the information in the **Annual US Precipitation** folder, what range of precipitation does your location fall in? What general direction would you need to travel to find locations with higher precipitation? Lower precipitation? Be sure to indicate your location in your response.

_____

_____

_____

_____

_____

_____

When you complete this part, turn off the **Annual US Precipitation** folder.

Instructions for Exploration 6.2 Precipitation Short Answer B:

1. Alternate turning on the *January 2011 Precipitation* and *June 2011 Precipitation* layers.

Exploration 6.2 Precipitation Short Answer B

B. What are the January and June precipitation levels in the selected desert and rainforest locations? Do precipitation levels vary at all between seasons, or do they remain consistent?

_____

_____

_____

_____

_____

_____

When you complete this exploration, collapse and turn off the **6.2 Precipitation** folder.

Exploration 6.3 – Clouds
Multiple Choice

Clouds are visible expressions of water vapor in the atmosphere that are made up of minute droplets of liquid water or ice crystals. Understanding cloud forms and cloud types can provide you with important clues to what is happening and what will happen in the atmosphere.

Instructions for all Parts:

1. Make sure you have opened the **KMZ** file from www.mygeoscienceplace.com

2. From the Places panel, expand **6. Atmospheric Moisture.kmz** and then open the **6.3 Clouds** folder.

Instructions for Exploration 6.3 Clouds Part A:

1. Double-click and turn on the **Cloud Height** folder.

2. The five labeled placemarks indicate possible heights of different cloud types. Double-click each placemark to zoom in to each one, noting the altitude at that location. Compare that altitude to the expected level of each cloud type.

Exploration 6.3 Clouds Part A

A. Based on altitude, which of the labeled placemarks in the **Cloud Heights** folder is in the correct placement for cirrus clouds?

1. A
2. B
3. C
4. D
5. E

When you complete this part, collapse and turn off the **Cloud Height** folder.

Instructions for Exploration 6.3 Clouds Part B:

1. Double-click the *Golden Gate Bridge* Gigapan photo.

2. Zoom in and examine the predominant cloud type found within this photo.

Exploration 6.3 Clouds Part B

B. What type of cloud is the most prevalent in the *Golden Gate Bridge* Gigapan photo?

1. Cumulonimbus
2. Stratus
3. Nimbostratus
4. Altocumulus
5. Cirrus

When you complete this part, turn off the *Golden Gate Bridge* Gigapan photo.

Instructions for Exploration 6.3 Clouds Part C:

1. Double-click the *View from Mendelpass* Gigapan photo.

2. Zoom in and examine the predominant cloud types found within this photo.

Exploration 6.3 Clouds Part C

C. What two types of cloud are most prevalent in the *View from Mendelpass* Gigapan photo?

1. Cirrocumulus and Altostratus
2. Altocumulus and Nimbostratus
3. Cirrostratus and Stratocumulus
4. Cirrus and Cumulus
5. Cirrocumulus and Cumulonimbus

When you complete this part, collapse and turn off the *View from Mendelpass* Gigapan photo.

Instructions for Exploration 6.3 Clouds Part D:

1. Double-click the *Pittsburgh* Gigapan photo.

2. Zoom in and examine the predominant cloud form found within this photo.

Exploration 6.3 Clouds Part D

D. What cloud form is the most prevalent in the *Pittsburgh* Gigapan photo?

1. Cirroform
2. Stratoform
3. Advection
4. Cumuloform
5. Radiation

When you complete this exploration, collapse and turn off the **6.3 Clouds** folder.

Exploration 6.3 – Clouds
Short Answer

Instructions for Exploration 6.3 Clouds Short Answer A:

1. Make sure you have opened the **KMZ** file from www.mygeoscienceplace.com

2. From the Places panel, expand **6. Atmospheric Moisture.kmz** and then open the **6.3 Clouds** folder.

3. Turn on and double-click the *Satellite Cloud Height* layer. This layer shows the recorded cloud heights over Central America measured in kilometers.

4. Turn on the *Satellite Cloud Temperature* layer. This layer shows the temperature of clouds over that same region, measured in degrees Kelvin.

5. Toggle between the two layers of cloud height and temperature over Central America, noting any relationship between the two variables in the cloud development at the time of the images.

Exploration 6.3 Clouds Short Answer A

A. Using the data in the *Satellite Cloud Height* and *Satellite Cloud Temperature* layers, describe the relationship between these two variables. Note examples in the cloud development from the images that support your claim and explain why this relationship exists.

_____

_____

_____

_____

_____

_____

_____

When you complete this part, collapse and turn off the **6.3 Clouds** folder.

Instructions for Exploration 6.3 Clouds Short Answer B:

1. Expand the *Weather* layer from the *Primary Database*.

2. Turn on the *Clouds* layer. Examine the global distribution of clouds, paying particular attention to regions of active cloud development.

Exploration 6.3 Clouds Short Answer B

B. Describe a region with high concentration of cloud development. You should indicate the physical location of this region as well as the potential reason for high cloud development in this region. Your answer should consider the seasonal difference that might be present from one part of the globe to another as well as address areas that may receive more precipitation than others.

_____

_____

_____

_____

_____

_____

_____

When you complete this exploration, collapse and turn off the *Weather* layer.

Exploration 6.4 – Acid Rain
Multiple Choice

Acidic materials can be deposited from the atmosphere via precipitation. Often called acid rain, anthropogenic sources of sulfur dioxides and nitrous oxides are the principle acidic constituents of this major hazard to the environment.

Instructions for all Parts:

1. Make sure you have opened the **KMZ** file from www.mygeoscienceplace.com

2. From the Places panel, expand **6. Atmospheric Moisture.kmz**.

3. Open the **6.4 Acid Rain** folder, then expand the **EPA Progress Reports** folder.

Instructions for Exploration 6.4 Acid Rain Parts A–C:

1. Open the *Maps* section. This folder contains data layers showing the deposition or concentration of significant components of acid rain, including sulfur dioxide, nitrates, and wet sulfates. Each compound includes data for 1989–1991 and for 2008–2010.

2. Examine the data for each of these acid rain–contributing compounds. Pay particular attention to where the significant impacts of these compounds are found and the trends between the 1989 and 2008 datasets for each.

Exploration 6.4 Acid Rain Part A

A. Based on the available data in the *Maps* section of the **EPA Progress Reports** folder, which of the following regions was most impacted by the various acid rain–producing compounds?

1. Upper Columbia River valley
2. The Gulf Coastal Plain
3. Ohio River valley
4. Lower Mississippi River valley
5. Arkansas River valley

Exploration 6.4 Acid Rain Part B

B. Based on the available data in the *Maps* section of the **EPA Progress Reports** folder, which of the following compounds saw the most widespread improvement (decrease) in concentration or deposition across the largest area from the 1989–1991 dataset to the 2008–2010 dataset?

1. Sulfur Dioxide
2. Ambient Sulfate
3. Ambient Nitrate
4. Wet Sulfate
5. Wet Inorganic Nitrogen

Exploration 6.4 Acid Rain Part C

C. Based on the available data in the *Maps* section of the **EPA Progress Reports** folder, which of the following compounds **did not** see significant *widespread* improvement (decrease) in concentration or deposition from the 1989–1991 dataset to the 2008–2010 dataset?

1. Sulfur Dioxide
2. Ambient Sulfate
3. Ambient Nitrate
4. Wet Sulfate
5. Wet Inorganic Nitrogen

When you complete the previous three parts, turn off and collapse the *Maps* section of the **EPA Progress Reports** folder.

Instructions for Exploration 6.4 Acid Rain Part D:

1. Open the *Point Emissions* section of the **EPA Progress Reports** folder. This folder contains point data for all facilities that have released a variety of atmospheric gasses during the period of 1990–2010. Data are shown for sulfur dioxide, nitrous oxides, carbon dioxides, and overall heat release. The point symbols for each layer are scaled based on the release of each component in 2010.

2. Turn on the *sulfur dioxide* (SO_2) layer to examine the release of this significant component of acid rain. Examine the areas of the country with the most concentrated levels of sulfur dioxide production.

Exploration 6.4 Acid Rain Part D

D. Based on the available data in the *Point Emissions* section of the **EPA Progress Reports** folder, which of the following states was home to the location with the greatest amount of sulfur dioxide (SO_2) emissions in 2010?

1. Ohio
2. Pennsylvania
3. Texas
4. West Virginia
5. Georgia

When you complete this exploration, collapse and turn off the **6.4 Acid Rain** folder.

Exploration 6.4 – Acid Rain
Short Answer

Instructions for all Parts:

1. Make sure you have opened the **KMZ** file from www.mygeoscienceplace.com

2. From the Places panel, expand **6. Atmospheric Moisture.kmz**.

3. Open the **6.4 Acid Rain** folder, then open the **EPA Progress Reports** folder.

Instructions for Exploration 6.4 Acid Rain Short Answer A:

1. Open the *Maps* section. This folder contains data layers showing the deposition or concentration of significant components of acid rain, including sulfur dioxide, nitrates, and wet sulfates. Each compound includes data for 1989–1991 and for 2008–2010.

2. Examine the data for each of the compounds provided in this folder. Note the changes in concentration and shifts in distribution of each comparing the 1989–1991 and 2008–2010 datasets.

Exploration 6.4 Acid Rain Short Answer A

A. Which of the compounds saw the least improvement in values across a large area from 1989–1991 to 2008–2010? What could explain the movement of high concentration areas and the lack of significant overall improvement in values for this compound?

_____
_____
_____
_____
_____
_____
_____

When you complete this part, turn off and collapse the *Maps* section of the **EPA Progress Reports** folder.

Instructions for Exploration 6.4 Acid Rain Short Answer B:

1. Open the *Point Emissions* section of the **EPA Progress Reports** folder. This folder contains point data for all facilities that have released a variety of atmospheric gasses during the period of 1990–2010. Data are shown for sulfur dioxide, nitrous oxides, carbon dioxides, and overall heat release. The point symbols for each layer are scaled based on the release of each component in 2010.

2. Fly to your location and identify the point location nearest you. Click on the closest facility point to open a chart displaying the relevant emissions data for this location for 1990–2010.

Exploration 6.4 Acid Rain Short Answer B

B. What is the name of the closest facility to your location? Briefly describe the emission trends that have occurred during the 20-year period shown in the data. Hypothesize reasons why the values have increased or decreased.

_____

_____

_____

_____

_____

_____

_____

When you complete this exploration, collapse and turn off the **6.4 Acid Rain** folder.

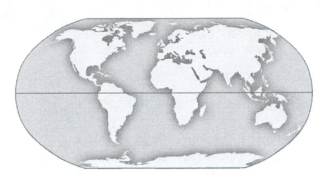

Encounter Physical Geography

Name:_____

Date: _____

Chapter 7
Atmospheric Disturbances

Exploration 7.1 – Air Masses and Fronts
Multiple Choice

The troposphere is not a uniform zone. Within this lower part of the atmosphere, there are regional bodies of air that have some degree of uniformity in temperature and humidity. These air masses move and interact along boundaries known as fronts, resulting in the changes in weather we experience on a daily basis.

Instructions for all Parts:

1. Make sure you have opened the **KMZ** file from www.mygeoscienceplace.com.

2. From the Places panel, expand **7. Atmospheric Disturbances.kmz** and then open the **7.1 Air Masses and Fronts** folder.

Instructions for Exploration 7.1 Air Masses and Fronts Parts A-B:

1. Turn on and double-click the **Source Regions** folder. The five labeled polygons in this folder represent potential source regions for typical air masses that are present over North America.

2. Examine each of the source region polygons paying attention to the type of air that would likely come from each.

Exploration 7.1 Air Masses and Fronts Part A

A. What location is the most likely source region for a maritime tropical air mass?

1. A
2. B
3. C
4. D
5. E

Exploration 7.1 Air Masses and Fronts Part B

B. What location is the most likely source region for a continental polar air mass?

1. A
2. B
3. C
4. D
5. E

When you complete the previous two parts, collapse and turn off the **Source Regions** folder.

Instructions for Exploration 7.1 Air Masses and Fronts Parts C-D:

1. Turn on and double-click the **Fronts** folder. This folder contains a static image of the Earth entitled *Blue Marble* along with five placemarked locations throughout North America.

2. Examine the location of the five placemarked locations compared to the visual evidence of surface weather patterns evidenced in the image of the Earth. Note the potential impacts of surface conditions at each location at the time of the image and immediately following.

Exploration 7.1 Air Masses and Fronts Part C

C. Which of the five locations placemarked in the **Fronts** folder would be the most likely to be impacted by a warm front immediately after this image was taken?

1. A
2. B
3. C
4. D
5. E

Exploration 7.1 Air Masses and Fronts Part D

D. Which of the five locations placemarked in the **Fronts** folder would be the most likely to be impacted by a cold front immediately after this image was taken?

1. A
2. B
3. C
4. D
5. E

When you complete this exploration, turn off and collapse the **7.1 Air Masses and Fronts** folder.

Exploration 7.1 – Air Masses and Fronts
Short Answers

Instructions for all Parts:

1. Make sure you have opened the **KMZ** file from www.mygeoscienceplace.com.

2. From the Places panel, expand **7. Atmospheric Disturbances.kmz** and then open the **7.1 Air Masses and Fronts** folder.

Instructions for Exploration 7.1 Air Masses and Fronts Short Answer A:

1. Open the **Fronts** folder then turn on the *Blue Marble* layer. This image shows a static view of the Earth and includes an abundance of evidence of weather patterns.

2. Examine the *Blue Marble* image noting the locations of significant atmospheric features such as the ITCZ and Sub-Tropical Highs. You may find it helpful to turn on the Grid from the View menu in your evaluation of the location of these features.

Exploration 7.1 Air Masses and Fronts Short Answer A

A. Was this image taken during the Northern Hemisphere summer or winter? Explain your response using specific examples from the image.

_____
_____
_____
_____
_____
_____
_____

When you complete this part, turn off and collapse the **Fronts** folder.

Instructions for Exploration 7.1 Air Masses and Fronts Short Answer B:

1. Turn on and double-click the **Source Regions** folder. The five labeled polygons in this folder represent sample source regions for typical air masses that are common over North America.

2. Examine each of the source region polygons and correlate their locations with the air mass types as described in your textbook.

3. Find your location and determine the dominant air mass types that influence the conditions there.

Exploration 7.1 Air Masses and Fronts Short Answer B

B. What air mass types are the dominant influences at your location? What are the typical characteristics of those air mass types? Where do these air masses originate and what are the seasonal variations that occur? What is the air mass that is present currently? Be sure to indicate your location in your response.

_____

_____

_____

_____

_____

_____

_____

When you complete this exploration, turn off and collapse the **7.1 Air Masses and Fronts** folder.

Exploration 7.2 – Mid-Latitude Cyclones & Anti-Cyclones
Multiple Choice

Atmospheric disturbances such as mid-latitude cyclones and anti-cyclones produce a range of weather phenomena, from storms to clear weather. These features are migratory and usually have a brief duration but the weather they produce is relatively predictable.

Instructions for all Parts:

1. Make sure you have opened the **KMZ** file from www.mygeoscienceplace.com.

2. From the Places panel, expand **7. Atmospheric Disturbances.kmz** and then open the **7.2 Mid-Latitude Cyclones & Anti-Cylcones** folder.

Instructions for Exploration 7.2 Mid-Latitude Cyclones & Anti-Cyclones Parts A-B:

1. Turn on the _Blue Marble_ layer then turn on the **Highs and Lows** folder. The image that appears is a static view of the Earth and includes an abundance of evidence of weather patterns. Included with this image are five placemarked locations labeled _A_ through _E_.

2. Examine each of the five placemarked locations in the **Highs and Lows** folder in context of the cloud patterns seen in the _Blue Marble_ image layer. Pay particular attention to the areas where high and low pressure systems may be present.

Exploration 7.2 Mid-Latitude Cyclones & Anti-Cyclones Part A

A. Which of the five locations placemarked in the **Highs and Lows** folder represents a clearly defined area of surface high pressure?

  1. A, B, & C
  2. B, D, & E
  3. A, C, & E
  4. C, D, & E
  5. A, D, & E

Exploration 7.2 Mid-Latitude Cyclones & Anti-Cyclones Part B

B. Which of the five locations placemarked in the **Highs and Lows** folder represents a clearly defined area of surface low pressure?

  1. B & D
  2. A & C
  3. B & E
  4. C & D
  5. A & E

When you complete the previous two parts, turn off the **Highs and Lows** folder and the *Blue Marble* image layer.

Instructions for Exploration 7.2 Mid-Latitude Cyclones & Anti-Cyclones Parts C-D:

  1. Turn on the **Hypothetical Pressure Systems** folder. The graphics in this folder indicate areas of surface high and low pressure as well as associated frontal boundaries.

  2. Examine the surface pressure and frontal locations being sure to consider the impacts these features would have on weather conditions in the United States.

Exploration 7.2 Mid-Latitude Cyclones & Anti-Cyclones Part C

C. Based on the locations of the High and Low pressure cells indicated in the **Hypothetical Pressure Systems** folder, what is the likely wind direction at Kansas City, Missouri?

  1. Northeast
  2. Southeast
  3. South
  4. Northwest
  5. North

Exploration 7.2 Mid-Latitude Cyclones & Anti-Cyclones Part D

D. Based on the frontal model presented in the **Hypothetical Pressure Systems** folder, which city is most likely to experience an intense rain event followed by a marked drop in temperatures over the next few hours?

1. Atlanta
2. Nashville
3. Charlotte
4. Little Rock
5. Knoxville

When you complete this exploration, turn off the **7.2 Mid-Latitude Cyclones & Anti-Cyclones** folder.

Exploration 7.2 – Mid-Latitude Cyclones & Anti-Cyclones
Short Answers

Instructions for Exploration 7.2 Mid-Latitude Cyclones & Anti-Cyclones Short Answer A:

1. Make sure you have opened the **KMZ** file from www.mygeoscienceplace.com.

2. From the Places panel, expand **7. Atmospheric Disturbances.kmz** and then open the **7.2 Mid-Latitude Cyclones & Anti-Cylcones** folder.

3. Turn on the **Hypothetical Pressure Systems** folder.

4. Examine the location of Savannah, Georgia, paying attention to the context of the frontal features shown.

Exploration 7.2 Mid-Latitude Cyclones & Anti-Cyclones Short Answer A

A. Based on typical frontal dynamics and the scenario shown, describe the weather conditions in Savannah over the next few days. Be sure to indicate the relative timing of significant changes in precipitation patterns and temperature.

_____

_____

_____

_____

_____

_____

_____

When you complete this part, turn off the **Hypothetical Pressure Systems** folder.

Instructions for Exploration 7.2 Mid-Latitude Cyclones & Anti-Cyclones Short Answer B:

1. Fly to your location using Google Earth™.

2. From the *Weather* layer in the *Primary Database*, turn on the *Clouds* layer. Use the information in this layer to evaluate the context of current surface weather features.

Exploration 7.2 Mid-Latitude Cyclones & Anti-Cyclones Short Answer B

B. Using the information in the *Clouds* layer as evidence, describe the current impacts of mid-latitude cyclones and anti-cyclones at your location yesterday, today, and tomorrow. Be sure to indicate both your location and the date and time you examined the data.

When you complete this exploration, collapse and turn off the **7.2 Mid-Latitude Cyclones & Anti-Cyclones** folder.

Exploration 7.3 – Tropical Disturbances
Multiple Choice

Intense tropical disturbances, known as hurricanes in North America, develop in the tropics and occasionally move into the mid-latitudes. These storms can have dramatic impacts with their intense rainfall, high winds, and associated flooding. Certain regions of the United States are more impacted by these seasonal storms than others.

Instructions for all Parts:

1. Make sure you have opened the **KMZ** file from www.mygeoscienceplace.com.

2. From the Places panel, expand **7. Atmospheric Disturbances.kmz** and then open the **7.3 Tropical Disturbances** folder.

Instructions for Exploration 7.3 Tropical Disturbances Part A:

1. Open the **Atlantic Basin Tropical Cyclones** folder.

2. Turn on and double-click the *Katrina Path* layer. This line represents the path of Hurricane Katrina, a major hurricane of the 2005 season.

Exploration 7.3 Tropical Disturbances Part A

A. Which of the following statements is best supported by the *Katrina Path* layer?

1. Katrina began as a tropical low in the Gulf of Mexico.
2. Katrina's first landfall occurred in the Mississippi Delta south of New Orleans.
3. Katrina made landfalls near two major American cities.
4. The path of Katrina suggests that it created a problem with torrential rain and flooding in the Central American nation of Honduras.
5. Katrina remained an organized tropical system as it moved over New England.

When you complete this part, turn off the *Katrina Path* layer.

Instructions for Exploration 7.3 Tropical Disturbances Part B:

1. Ensure that the **Atlantic Basin Tropical Cyclones** folder is open then turn on and double-click the **2005** folder. In this folder, each tropical system from the 2005 season is represented by a single path. Each line contains information about the storms at various points along their paths. You might consider turning on one storm at a time to evaluate the paths of each

Exploration 7.3 Tropical Disturbances Part B

B. Based on the paths seen in the **2005** folder, how many unique tropical systems made landfall in the United States in 2005?

1. 0
2. 3 to 5
3. 6 to 9
4. 10 to 13
5. over 13

When you complete this part, turn off and collapse the **2005** folder.

Instructions for Exploration 7.3 Tropical Disturbances Part C:

1. Ensure that the **Atlantic Basin Tropical Cyclones** folder is open then turn on and double-click the **2011** folder. In this folder, each tropical system from the 2011 season is represented by a single path. The color of the line indicates the strength of each storm segment.

2. Click on the tropical system symbols along the storm paths to see information about each storm at those locations.

3. Evaluate the paths of the 2011 Atlantic Hurricane season focusing your attention on the storms which reached the highest level of intensity.

Exploration 7.3 Tropical Disturbances Part C

C. Based on the data in the **2011** folder, what was the minimum pressure in millibars that occurred in a tropical system during the 2011 season?

1. 938
2. 940
3. 942
4. 944
5. 946

When you complete this part, turn off and collapse the **2011** folder.

Instructions for Exploration 7.3 Tropical Disturbances Part D:

1. Turn on the **Tropical Storm Allison Satellite Animation** folder. This folder contains a time-lapse video showing the movement of Tropical Storm Allison, a significant event from the 2001 Atlantic Hurricane season.

2. Using the time slider in the top left corner of the 3D Viewer, view the animation of Tropical Storm Allison from June 2001. The imagery comes from NASA; it may take a few moments for individual images to load.

3. Evaluate the path of Tropical Storm Allison.

Exploration 7.3 Tropical Disturbances Part D

D. Which statement is most strongly supported by the animation shown in the *Tropical Storm Allison Satellite Animation* video layer?

1. Tropical Storm Allison's maximum latitude was 35° North.
2. The initial formation of Tropical Storm Allison occurred over Hispaniola.
3. Tropical Storm Allison made only one landfall in the United States.
4. The center of Tropical Storm Allison passed over more than five states.
5. Tropical Storm Allison caused major flooding in St. Louis, Missouri.

When you complete this exploration, collapse and turn off the **7.3 Tropical Disturbances** folder.

Exploration 7.3 – Tropical Disturbances
Short Answers

Instructions for all Parts:

1. Make sure you have opened the **KMZ** file from www.mygeoscienceplace.com.

2. From the Places panel, expand **7. Atmospheric Disturbances.kmz** and then open the **7.3 Tropical Disturbances** folder.

Instructions for Exploration 7.3 Tropical Disturbances Short Answer A:

1. Turn on and open the **Hurricane Sandy** folder. This folder contains several layers related to the landfall of this significant storm which impacted the northeastern United States in late October 2012. The *Track* layer indicates the generalized path of the center of Sandy as it moved up the coast and then progressed inland. The **Mantoloking, New Jersey** and **Ocean City, Maryland** folders contain imagery of damage along the coast at two locations. The Ocean City imagery was captured on October 31, 2012; the Mantoloking imagery was captured on November 1, 2012. Imagery from both locations was collected by the National Oceanic and Atmospheric Administration's National Geodetic Survey.

2. Note the relative location of the two indicated communities with respect to the path of the storm.

3. Zoom in and explore the landscapes shown at both Mantoloking and Ocean City, noting significant impacts of Hurricane Sandy. Be sure to turn off the post-storm imagery and use the Historic Imagery time slider or default Google Earth™ imagery to examine these locations as they looked prior to the storm.

Exploration 7.3 Tropical Disturbances Short Answer A

A. Summarize the impacts of Hurricane Sandy at the two locations. In addition to describing the damage, you should indicate a reason for the disparity of impacts between these two sites. It may be helpful for you to consider the dynamics of tropical cyclones and the quadrant of a typical storm that receives the brunt of the storm effects. Consider the relative location of Sandy's path to each of the two locations.

_____

_____

_____

_____

_____

_____

_____

When you complete this part, turn off the **Hurricane Sandy** folder.

Instructions for Exploration 7.3 Tropical Disturbances Short Answer B:

1. Open the **Atlantic Basin Tropical Cyclones** folder.

2. In turn, examine the storm paths for the 2005 and 2011 hurricane seasons by turning each folder on and off. Click on individual storm paths to find out relevant information for each system. (NOTE: Because of the amount of data in each folder, you may find it easier to have only one season turned on at a time.)

3. Examine the patterns for the named tropical disturbances in the 2005 and 2011 seasons. Pay particular attention to the origins and destinations of storms, frequency of tropical events, and impacts on the Caribbean and mainland North America.

Exploration 7.3 Tropical Disturbances Short Answer B

B. Compare and contrast the 2005 and 2011 Atlantic Hurricane seasons. In your evaluation, be sure to indicate the approximate number of storms in each season, the common areas of storm cyclogenesis, relative intensity and tracks of major storms, and direct impacts on land areas.

_____

_____

_____

_____

_____

_____

_____

When you complete this exploration, collapse and turn off the **7.3 Tropical Disturbances** folder.

7.4 – Localized Severe Weather
Multiple Choice

Atmospheric disturbances can occur on a much smaller scale than tropical cyclones but are nonetheless destructive and impact landscapes on a local scale. These include phenomena such as thunderstorms and tornadoes.

Instructions for all Parts:

1. Make sure you have opened the **KMZ** file from www.mygeoscienceplace.com.

2. From the Places panel, expand **7. Atmospheric Disturbances.kmz** and then open the **7.4 Localized Severe Weather** folder.

Instructions for Exploration 7.4 Localized Severe Weather Parts A-B:

1. Turn on and double-click the *Thunderstorms* layer. This image shows a line of severe thunderstorms stretching across the central United States. The storms are indicated by the large masses of clouds along and ahead of a distinct frontal boundary.

2. Examine the imagery of these severe storms being sure to consider the impact this weather system would have on the regions immediately behind and ahead of the frontal boundary.

Exploration 7.4 Localized Severe Weather Part A

A. Based on the evidence in the *Thunderstorms* image layer, which of the following cities is most likely to be experiencing a thunderstorm at the moment?

1. Cedar Rapids
2. Omaha
3. Kansas City
4. St. Louis
5. Milwaukee

Exploration 7.4 Localized Severe Weather Part B

B. Based on the evidence in the *Thunderstorms* image layer, which of the following cities is most likely to experience a thunderstorm in the next 6 to 12 hours?

1. Kansas City
2. St. Louis
3. Omaha
4. Wichita
5. Des Moines

When you complete the previous two parts, turn off the *Thunderstorm* layer.

Instructions for Exploration 7.4 Localized Severe Weather Part C:

1. Double-click the *Joplin Tornado* placemark to zoom into the city of Joplin, Missouri.

2. On May 22, 2011, a large tornado struck the city of Joplin. The damage path of this storm is seen running across the center of the image. The most significant damage is in the center of the path, where many structures were completely destroyed; structures farther from the center of the damage path were less severely impacted.

3. Examine the imagery of the city of Joplin, paying particular attention to the damage seen along the path of the 2011 tornado.

Exploration 7.4 Localized Severe Weather Part C

C. Which of the following statements is best supported by the evidence seen in the imagery at the *Joplin Tornado* placemark?

1. The area impacted had an exceptionally high number of swimming pools, as evidenced by the blue polygons throughout the affected neighborhoods.
2. The town's largest high school escaped significant damage during the storm.
3. In some locations the width of the damage path exceeds 1 km.
4. The town's largest hospital escaped significant damage.
5. The tornado only moved in a southwest to northeast direction as it moved through the city.

Instructions for Exploration 7.4 Localized Severe Weather Part D:

1. Double-click the *Tuscaloosa Tornado* placemark to zoom to an area southwest of the city of Tuscaloosa, Alabama. On April 27, 2011, a large tornado touched down near this location and moved northeast through Tuscaloosa and towards Birmingham.

2. Evaluate the imagery of the Tuscaloosa tornado damage path paying particular attention to the total linear extent of the damage.

Exploration 7.4 Localized Severe Weather Part D

D. Based on the Google Earth™ imagery in the *Tuscaloosa Tornado* layer, what is the approximate length of the destruction path of the 2011 Tuscaloosa Tornado?

1. 15 km
2. 25 km
3. 65 km
4. 100 km
5. 150 km

When you complete this exploration, collapse and turn off the **7.4 Localized Severe Weather** folder.

7.4 – Localized Severe Weather
Short Answers

Instructions for all Parts:

1. Make sure you have opened the **KMZ** file from www.mygeoscienceplace.com.

2. From the Places panel, expand **7. Atmospheric Disturbances.kmz** and then open the **7.4 Localized Severe Weather** folder.

Instructions for Exploration 7.4 Localized Severe Weather Short Answer A

1. Double-click the *Tuscaloosa Tornado* placemark. This location represents the approximate starting location for a major tornado that struck this area in April 2011.

2. Double-click the *Joplin Tornado* placemark. The city of Joplin was struck by a major tornado in May 2011.

3. Examine the evidence of damage of the storms in Tuscaloosa and Joplin using imagery in Google Earth™. Be sure to use historical imagery if appropriate.

4. Use an outside source to research the impacts of both storms on their respective cities.

Exploration 7.4 Localized Severe Weather Short Answer A

A. How many fatalities were caused by the 2011 tornadoes in Tuscaloosa and Joplin? Based on the evidence in the imagery seen in the damage paths associated with the *Tuscaloosa Tornado* and *Joplin Tornado* placemarks, as well as any appropriate outside sources, create a hypothesis for why one storm was more deadly than the other. Be sure to consider factors such as the difference in size and length of damage paths as seen in the Google Earth™ imagery.

_____

_____

_____

_____

_____

_____

IMPORTANT: When you complete this part, be sure the Historical Imagery slider is turned off.

Instructions for Exploration 7.4 Localized Severe Weather Short Answer B:

1. Double-click and turn on the **Wildland Fire Assessment System (WFAS)** folder. This data, collected and disseminated by the National Wildfire Coordinating Group, provides daily wildfire risk assessments for selected forecasting stations across the United States. Each station is represented by a color coded placemark where green represents relatively low fire danger and red relatively high fire danger.

2. Click on any forecasting station placemark to open a window containing relevant observed and forecast data associated with wildfire risk, including weather and fuel factors.

3. From the *Primary Database*, turn on any layers from the *Weather* folder that might help show the current surface weather conditions.

Exploration 7.4 Localized Severe Weather Short Answer B

B. Using information from the **Wildland Fire Assessment System** folder, surface weather conditions, and evidence of local vegetation patterns in the Google Earth™ imagery, determine what areas of the United States are currently at the greatest risk for a lightning generated wildfire. Indicate the name of WFAS forecast stations near this location and provide a summary of the risk factors that contribute to the high risk classification.

_____

_____

_____

_____

_____

_____

When you complete this exploration, collapse and turn off the **7.4 Localized Severe Weather** folder.

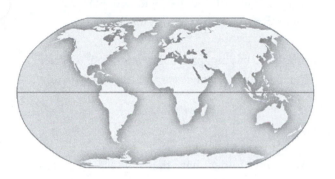

Encounter Physical Geography

Name:_____

Date: _____

Chapter 8
Climate and Climate Change

Exploration 8.1 – Climate Classification Types I
Multiple Choice

The study of climate begins with the attempt to classify the distribution of global climate patterns. Beginning with the Greeks' rudimentary classification system and moving to the more complex systems utilized today, such as the Köppen system, humans have attempted to simplify the complex multi-variable phenomenon known as climate. One way to do this is through the construction of simple charts, or climographs, that illustrate temperature and precipitation at a specific site throughout the year.

Instructions for all Parts:

1. Make sure you have opened the **KMZ** file from www.mygeoscienceplace.com.

2. From the Places panel, expand **8. Climate & Climate Change.kmz** and then open the **8.1 Climate Classification Types I** folder.

Instructions for Exploration 8.1 Climate Classification Types I Part A:

1. Use Google Earth™ to locate the following five African cities: Algiers, Abuja, Cape Town, Rabat, and Tunis. Be certain to note the latitude of each city.

Exploration 8.1 Climate Classification Types I Part A

A. Based on their locations relative to significant latitudinal regions, which of the following cities would likely fall within the region once defined as the "Torrid Zone" by the Greeks?

1. Cape Town
2. Rabat
3. Abuja
4. Tunis
5. Algiers

Instructions for Exploration 8.1 Climate Classification Types I Part B:

1. Open and turn on the *Extreme Locations* folder. Each of the five labeled placemarks represents locations around the globe where an extreme weather record has been set.

2. Examine the five placemarks labeled *A* through *E*, paying particular attention to their location relative to recorded historical weather extremes. If necessary, consult your text or other outside sources.

Exploration 8.1 Climate Classification Types I Part B

B. Which of the placemarks in the *Extreme Locations* folder has the highest recorded temperature?

1. A
2. B
3. C
4. D
5. E

When you complete the previous part, close and turn off the **Extreme Locations** folder.

Instructions for Exploration 8.1 Climate Classification Types I Parts C–D:

1. Open the **Climograph** folder. This folder contains two sample climographs. Each represents the temperature and precipitation averages for a location in North America.

2. Toggle between *Climograph 1* and *Climograph 2* and consider the locations that would have similar temperature and precipitation patterns.

Exploration 8.1 Climate Classification Types I Part C

C. Which of the following cities is represented by the data presented in *Climograph 1*?

1. Winnipeg, Manitoba
2. Miami, Florida
3. San Francisco, California
4. Phoenix, Arizona
5. Boston, Massachusetts

Exploration 8.1 Climate Classification Types I Part D

D. Which of the following cities is represented by the data presented in *Climograph 2*?

1. Winnipeg, Manitoba
2. Miami, Florida
3. San Francisco, California
4. Phoenix, Arizona
5. Boston, Massachusetts

When you complete this exploration, turn off and collapse the **8.1 Climate Classification Types I** folder.

Exploration 8.1 – Climate Classification Types I
Short Answer

Instructions for all Parts:

1. Make sure you have opened the **KMZ** file from www.mygeoscienceplace.com.

2. From the Places panel, expand **8. Climate & Climate Change.kmz** and then open the **8.1 Climate Classification Types I** folder.

Instructions for Exploration 8.1 Climate Classification Types I Short Answer A:

1. Open the **Climograph** folder. This folder contains two sample climographs. Each represents the temperature and precipitation averages for a location in North America.

2. Toggle between *Climograph 1* and *Climograph 2* to examine the climate patterns described for these locations.

Exploration 8.1 Climate Classification Types I Short Answer A

A. Select one of the two climographs from the **Climograph** folder. Note the expected similarities and differences between this climograph and a theoretical one for your location. Be sure to explain why the differences exist and note your location.

_____
_____
_____
_____
_____
_____
_____

When you complete this part, turn off and collapse the **Climograph** folder.

Instructions for Exploration 8.1 Climate Classification Types I Short Answer B:

1. Open and turn on the **Extreme Locations** folder. Each of the five labeled placemarks represents locations around the globe where an extreme weather record has been set.

Exploration 8.1 Climate Classification Types I Short Answer B

B. Select two of the placemarked locations from the *Extreme Locations* folder. Note the climate classification at those locations and research the weather extreme that was recorded at each. Based on your knowledge of climate, explain why those extreme recordings were more likely at those places.

_____

_____

_____

_____

_____

_____

When you complete this exploration, turn off and collapse the **8.1 Climate Classification Types I** folder.

Exploration 8.2 – Climate Classification Types II
Multiple Choice

Most climate classification systems emphasize variations in precipitation and temperature to develop their schemes. The Köppen climate classification system uses average annual and average monthly values of temperature and precipitation as its basis for classification. The system divides the world's climates into six major categories and numerous sub-categories that result from specific patterns of seasonality and extremes in temperature and precipitation.

Instructions for all Parts:

1. Make sure you have opened the **KMZ** file from www.mygeoscienceplace.com.

2. From the Places panel, expand **8. Climate & Climate Change.kmz** and then open the **8.2 Climate Classification Types II** folder. The polygons in this folder represent selected climate zones based on the Köppen climate classification system.

3. Turn on each of the *Climate* polygons labeled *Climate 1* through *Climate 4* and note the distribution of each climate zone.

Exploration 8.2 Climate Classification Types II Part A

A. What climate zone is represented by the *Climate 1* polygon?

1. Humid Continental
2. Tropical Wet
3. Mediterranean
4. Midlatitude Steppe
5. Sub-Arctic

Exploration 8.2 Climate Classification Types II Part B

B. What climate zone is represented by the *Climate 2* polygon?

1. Humid Continental
2. Tropical Wet
3. Mediterranean
4. Midlatitude Steppe
5. Sub-Arctic

Exploration 8.2 Climate Classification Types II Part C

C. What climate zone is represented by the *Climate 3* polygon?

1. Humid Continental
2. Tropical Wet
3. Mediterranean
4. Midlatitude Steppe
5. Sub-Arctic

Exploration 8.2 Climate Classification Types II Part D

D. What climate zone is represented by the *Climate 4* polygon?

1. Humid Continental
2. Tropical Wet
3. Mediterranean
4. Midlatitude Steppe
5. Sub-Arctic

When you complete this exploration, turn off and collapse the **8.2 Climate Classification Types II** folder.

Exploration 8.2 – Climate Classification Types II
Short Answer

Instructions for Exploration 8.2 Climate Classification Types II Short Answer A:

1. Use your textbook and Google Earth™ to identify a major world city that is located in a tropical monsoon climate.

Exploration 8.2 Climate Classification Types II Short Answer A

A. Indicate the name and location of the city you've selected. Explain the physical landscape of the region and then describe the social and cultural impacts this type of climate might have on a city of this size.

_____

_____

_____

_____

_____

_____

_____

Instructions for Exploration 8.2 Climate Classification Types II Short Answer B:

1. Navigate to your location in Google Earth™. Examine the physical landscape found in your region, paying particular attention to the patterns found in the local vegetation.

2. Use your textbook to identify another location on a different continent that shares the same climate classification as your location. Examine the physical landscape and vegetation patterns of that location.

Exploration 8.2 Climate Classification Types II Short Answer B

B. Describe the similarities and differences in terrain and vegetation patterns that exist between your location and another location on a different continent that shares the same climate classification. Be sure to indicate the location of your city as well as the second location using latitude and longitude.

_____

_____

_____

_____

_____

_____

_____

When you complete this exploration, turn off and collapse the **8.2 Climate Classification Types II** folder.

Exploration 8.3 – Global Climate Change Projections
Multiple Choice

Detailed analysis of paleoclimates through techniques of dendrochronology, ice cores, oxygen isotope analysis, and pollen analysis have helped us understand the cyclical and dynamic nature of Earth's climate. These studies partnered with analysis of present-day data related principally to greenhouse gasses allow us to make projections as to climate trends over the next century.

Instructions for all Parts:

1. Make sure you have opened the **KMZ** file from www.mygeoscienceplace.com.

2. From the Places panel, expand **8. Climate & Climate Change.kmz** and then open the **8.3 Global Climate Change Projections** folder.

Instructions for Exploration 8.3 Global Climate Change Projections Part A:

1. Open the **Medium Emissions Scenario** folder then open the **Average Temperature Change** folder. This folder contains global average temperature change scenarios projected at 30-year intervals throughout the 21st century for a scenario that assumes a medium rate of atmospheric emissions.

2. Examine each of the four scenario layers for the *2000s, 2030s, 2060s*, and *2090s*, paying attention to the regions of the world where temperature changes during the 21st century are the greatest.

Exploration 8.3 Global Climate Change Projections Part A

A. Which of the following locations would see the greatest amount of average temperature change by the 2090s?

1. Antarctica
2. South America
3. Sub-Saharan Africa
4. the Arctic
5. Southeast Asia

When you complete this question, collapse the **Average Temperature Change** folder.

Instructions for Exploration 8.3 Global Climate Change Projections Part B:

1. Ensure that the **Medium Emissions Scenario** folder is open.

2. Open the **Average Precipitation Change** folder. This folder contains global average precipitation change scenarios projected at 30-year intervals throughout the 21st century for a scenario that assumes a medium rate of atmospheric emissions.

3. Examine each of the four scenario layers for the *2000s, 2030s, 2060s*, and *2090s*. Be sure to note the relationship between predicted precipitation changes and the location of major global circulation components.

Exploration 8.3 Global Climate Change Projections Part B

B. What global surface component of pressure and wind correlates with the area expected to see the greatest increase in precipitation by 2090?

1. ITCZ
2. Westerlies
3. Sub-Tropical Highs
4. Polar Easterlies
5. Southwest Trade Winds

When you complete the previous two parts, collapse and turn off the **Medium Emissions Scenario** folder.

Instructions for Exploration 8.3 Global Climate Change Projections Part C:

1. Open the **High Emissions Scenario** folder then open the **Average Temperature Change** folder. This folder contains global average temperature change scenarios projected at 30-year intervals throughout the 21st century for a scenario that assumes a high rate of atmospheric emissions.

2. Examine each of the four scenario layers for the *2000s, 2030s, 2060s*, and *2090s*.

Exploration 8.3 Global Climate Change Projections Part C

C. Based on the data in the **Average Temperature Change** folder within the **High Emissions Scenario**, which of the following statements is most strongly supported?

1. South America will exhibit the most widespread increase in temperature.
2. Antarctica is projected to cool slightly.
3. The mid-latitude landmasses are projected to increase in temperature more than the low-latitude landmasses.
4. The islands of Southeast Asia have an average temperature increase of greater than 8° C.
5. The Sahara Desert is projected to see the greatest increase in temperature.

When you complete this part, turn off and collapse the **Average Temperature Change** folder.

Instructions for Exploration 8.3 Global Climate Change Projections Part D:

1. Ensure that the **High Emissions Scenario** folder is open.

2. Turn on and open the **Average Precipitation Change** folder. This folder contains global average precipitation change scenarios projected at 30-year intervals throughout the 21st century for a scenario that assumes a high rate of atmospheric emissions.

3. Examine each of the four scenario layers for the *2000s, 2030s, 2060s*, and *2090s*.

Exploration 8.3 Global Climate Change Projections Part D

D. Which of the following important agricultural regions will likely experience the greatest decrease in average precipitation in the high emissions scenario?

1. Northern European Plain
2. Deccan Plateau
3. Southeast Australia
4. American Southern Plains
5. Pampas

When you complete this exploration, collapse and turn off the **8.3 Global Climate Change Projections** folder.

Exploration 8.3 – Global Climate Change Projections
Short Answer

Instructions for all Parts:

1. Make sure you have opened the **KMZ** file from www.mygeoscienceplace.com.

2. From the Places panel, expand **8. Climate & Climate Change.kmz** and then open the **8.3 Global Climate Change Projections** folder.

Instructions for Exploration 8.3 Global Climate Change Projections Short Answer A:

1. Open the **Medium Emissions Scenario** folder then open the **Average Temperature Change** folder. This folder contains global average temperature change scenarios projected at 30-year intervals throughout the 21st century for a scenario that assumes a medium rate of atmospheric emissions.

2. Examine each of the four scenario layers for the *2000s, 2030s, 2060s*, and *2090s*. Identify the world region that is projected to experience the greatest increase in temperature during the 21st century.

Exploration 8.3 Global Climate Change Projections Short Answer A

A. What region is expected to see the greatest increase in temperatures during the 21st century? What characteristics of this region make projected temperature changes more pronounced and what impacts will that have on the region's physical environment?

_____

_____

_____

_____

_____

_____

_____

When you complete this part, collapse and turn off the **Medium Emissions Scenario** folder.

Instructions for Exploration 8.3 Global Climate Change Projections Short Answer B:

1. Open the **Low Emissions Scenario** folder then open the **Average Precipitation Change** folder. Turn on the *2090s* layer.

2. Examine the potential changes in global precipitation in the low emissions scenario for the 2090s.

3. Open the **High Emissions Scenario** folder then open the **Average Precipitation Change** folder. Turn on the *2090s* layer.

4. Compare the precipitation projections for the 2090s for both the Low Emissions and High Emissions scenarios.

Exploration 8.3 Global Climate Change Projections Short Answer B

B. Identify one location with a significant discrepancy between the projections for the High Emissions and Low Emissions scenarios. What geographic variables may make this location more likely to see a variable impact in precipitation change due to emissions?

_____

_____

_____

_____

_____

_____

_____

When you complete this exploration, collapse and turn off the **8.3 Global Climate Change Projections** folder.

Exploration 8.4 – Global Climate Change Impacts
Multiple Choice

Climate change has the potential to impact all parts of the Earth but those impacts will likely be considerably different from one region to another. In some locations, increases in temperature or precipitation may be a boon for agriculture while in others, the spread of disease or increased flooding may pose a greater risk to life and property.

Instructions for all Parts:

1. Make sure you have opened the **KMZ** file from www.mygeoscienceplace.com.

2. From the Places panel, expand **8. Climate & Climate Change.kmz** and then open the **8.4 Global Climate Change Impacts** folder.

Instructions for Exploration 8.4 Global Climate Change Impacts Part A:

1. Open the **Global Food Security Projections** folder. The focus on most climate change studies is on the negative impacts of change; however, for agricultural production some regions will see positive increases in criteria such as growing season length and crop suitability. The data folders contained here illustrate some of these potential positive impacts.

2. Double-click the **Length of Growing Season** folder.

3. Alternate turning on the **Reference: 1961–1990** and **2080s** folders to examine the recent historical average and projection for growing season length around the world. Note the patterns of change seen across the North American continent.

Exploration 8.4 Global Climate Change Impacts Part A

A. Which North American region is projected to see the greatest increase in growing season length by 2080?

1. Mojave Desert
2. Quebec
3. Florida Panhandle
4. Great Plains
5. Central Rocky Mountain

When you complete this part, collapse and turn off the **Global Food Security Projections** folder.

Instructions for Exploration 8.4 Global Climate Change Impacts Part B:

1. Open the **Malaria in Africa** folder. The spread of malaria is facilitated by warm, wet conditions, which increase the breeding of mosquitos. The data in this folder shows potential for the spread of malaria in Africa over the 21st century.

2. Turn on and double-click the **Malaria** folder. The individual folder layers contained here show the historical baseline and predicted potential spread of malaria for decades throughout the 21st century. The **Time Animation** folder also contained in the **Malaria in Africa** folder displays the same data using the Historical Imagery time slider.

3. Use the **Malaria in Africa** datasets to evaluate the changes in the potential spread of malaria over the course of the 21st century.

Exploration 8.4 Global Climate Change Impacts Part B

B. Which of the following statements is best supported by the information in the **Malaria in Africa** folder?

1. Malaria transmission will increasingly occur at higher elevation locations.
2. Malaria is not found in coastal areas of Africa.
3. The area of the longest malarial transmission period is centered on the Tropic of Capricorn.
4. By 2070, all countries on the African continent will exhibit areas suitable for malaria transmission.
5. Somalia has been and will remain a focal point for malaria transmission.

When you complete this part, collapse and turn off the **Malaria in Africa** folder.

Instructions for Exploration 8.4 Global Climate Change Impacts Part C:

1. Turn on the **Runoff Projection** folder. This data shows the projection for relative change in runoff between early in this century through the 2080s.

2. Evaluate the data shown in this folder, making note of how these changes may affect rivers around the world.

Exploration 8.4 Global Climate Change Impacts Part C

C. Based on the data in the **Runoff Projection** folder, which of the following rivers would likely see the biggest decrease in runoff by 2080?

1. St. Laurence River
2. Colorado River
3. Brahmaputra River
4. Amur River
5. Yangtze River

When you complete this part, turn off the **Runoff Projection** folder.

Instructions for Exploration 8.4 Global Climate Change Impacts Part D:

1. Turn on the **Deaths from Climate Change** folder. This data shows the estimated number of deaths attributable to climate change impacts by country in the year 2000.

2. Evaluate the patterns of deaths attributable to climate change, noting the presence of any regional patterns.

Exploration 8.4 Global Climate Change Impacts Part D

D. According to the data in the **Deaths from Climate Change** folder, what world region has the highest death rates attributable to climate change?

1. Europe
2. North America
3. South Asia
4. Sub-Saharan Africa
5. Oceania

When you complete this exploration, collapse and turn off the **8.4 Global Climate Change Impacts** folder.

Exploration 8.4 – Global Climate Change Impacts
Short Answer

Instructions for all Parts:

1. Make sure you have opened the **KMZ** file from www.mygeoscienceplace.com.

2. From the Places panel, expand **8. Climate & Climate Change**.**kmz** and then open the **8.4 Global Climate Change Impacts** folder.

Instructions for Exploration 8.4 Global Climate Change Impacts Short Answer A:

1. Turn on the **Deaths from Climate Change** folder. This data shows the estimated number of deaths attributable to climate change impacts by country in the year 2000.

2. Evaluate the patterns of deaths attributable to climate change, noting the presence of any regional patterns.

Exploration 8.4 Global Climate Change Impacts Short Answer A

A. What factors contribute to the high rates of climate deaths in some regions of the world? How do you think the attitudes of Americans toward the issue of climate change would be different if the data forecasted greater health security risks in North America?

_____

_____

_____

_____

_____

_____

_____

When you complete this part, turn off the **Deaths from Climate Change** folder.

Instructions for Exploration 8.4 Global Climate Change Impacts Short Answer B:

1. Turn on and double-click the **Sea Level Rise** folder. This data shows the impacts a 4 meter rise in sea level would have on the south Asian nation of Bangladesh.

2. Using outside sources and any evidence from the Google Earth™ database, research the nation of Bangladesh to determine the social and economic characteristics of its population.

Exploration 8.4 Global Climate Change Impacts Short Answer B

B. Briefly describe the spatial extent of the impacts of a 4 meter rise in sea level on Bangladesh. Based on your research, what impact would this change have on the people of this nation in terms of economic livelihood and security?

_____

_____

_____

_____

_____

_____

_____

When you complete this exploration, collapse and turn off the **8.4 Global Climate Change Impacts** folder.

Encounter Physical Geography

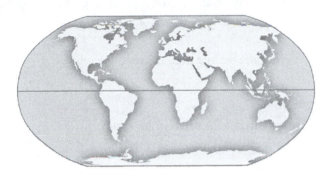

Name:_____

Date: _____

Chapter 9
The Hydrosphere

Exploration 9.1 – Ocean Waters
Multiple Choice

Oceans cover nearly three-quarters of the surface of the planet. Viewed from space, it is easy to consider these vast bodies of water as uniform features. In reality, however, there are tremendous variations in depth, temperature, and circulation patterns throughout the oceans. Increasingly, human activity is impacting even the most remote locations.

Instructions for all Parts:

1. Make sure you have opened the **KMZ** file from www.mygeoscienceplace.com.

2. From the Places panel, expand **9. The Hydrosphere.kmz** and then open the **9.1 Ocean Waters** folder.

Instructions for Exploration 9.1 Ocean Waters Part A:

1. Turn on the **Depth** folder. This map overlay displays the bathymetry of the ocean floors, showing the depth below sea level in meters.

2. Evaluate the overlay of ocean depth noting the general pattern of depth in major ocean basins.

Exploration 9.1 Ocean Waters Part A

A. Which of the following statements is best supported by the evidence in the **Depth** folder?

1. The Arctic Ocean is generally the shallowest of the world's four major ocean basins.
2. The western Pacific Ocean uniformly exceeds 8000 m in depth.
3. The continental shelves of both North and South America are generally more extensive on the western edge of the continents than on the east.
4. The North and South Atlantic Oceans are deepest along the mid-oceanic ridge.
5. Amongst the Indonesian island chain, the deepest ocean depths are found between Borneo and Sumatra.

When you complete this part, turn off the **Depth** folder.

Instructions for Exploration 9.1 Ocean Waters Part B:

1. Turn on the **Current Labels** folder. This overlay displays a generalized pattern of ocean currents around the globe.

2. Examine the pattern of ocean circulation considering the source area for each current.

Exploration 9.1 Ocean Waters Part B

B. Based on the general circulation pattern shown in the **Current Labels** folder, which of the following currents is most likely a cold-water current?

1. Equatorial Countercurrent
2. West Australia
3. Gulf Stream
4. East Australia
5. Brazil

When you complete the previous part, turn off the **Current Labels** folder.

Instructions for Exploration 9.1 Ocean Waters Parts C–D:

1. Turn on the **Human Impacts** folder. This overlay displays the impact of a multitude of variables that contribute to impacts on the world's ocean. The most heavily impacted areas are depicted in orange and red, and the least impacted areas are depicted in blue.

2. Examine the pattern of impacts on the world's oceans, being sure to note areas of high human impact.

Exploration 9.1 Ocean Waters Part C

C. Which of the following major bodies of water has the highest human impact, as indicated by the data in the **Human Impacts** folder?

1. Bay of Bengal
2. Black Sea
3. Hudson Bay
4. East China Sea
5. Red Sea

When one studies the Human Impact overlay, the presence of specific human activity, such as energy development, travel and tourism, and fishing, are evident. For instance, highly traveled shipping routes are visible as streaks of higher impact zones between major ports and passages.

Exploration 9.1 Ocean Waters Part D

D. Based on the data in the **Human Impacts** folder, which of the following statements is **not** supported?

1. Significant shipping traffic moves from Puerto Rico to both the southeastern United States and Europe.
2. The northeast coast of the United States exhibits the most widespread impact when compared with other US coastlines.
3. In terms of impacts near Italy, the Tyrrhenian Sea has relatively higher levels of impacts compared to the Adriatic.
4. The North Sea has the highest human impacts of European waters.
5. The exceptionally high levels of manufacturing in East Asia can be found in the impacts of neighboring water bodies.

When you complete this exploration, turn off and collapse the **9.1 Ocean Waters** folder.

Exploration 9.1 – Ocean Waters
Short Answer

Instructions for all Parts:

1. Make sure you have opened the **KMZ** file from www.mygeoscienceplace.com.

2. From the Places panel, expand **9. The Hydrosphere.kmz** and then open the **9.1 Ocean Waters** folder.

Instructions for Exploration 9.1 Ocean Waters Short Answer A:

1. Turn on the **Human Impacts** folder. This overlay displays the impact of a multitude of variables that contribute to impacts on the world's ocean. The most heavily impacted areas are depicted in orange and red, and the least impacted areas are depicted in blue.

2. Note that in a number of locations around the globe, highly traveled shipping routes are visible as streaks of higher impact zones between major ports and passages.

Exploration 9.1 Ocean Waters Short Answer A

A. Identify the latitude and longitude of two instances of shipping lanes. Hypothesize what type of activity account for the high shipping volume in these locations. Be sure to consider the products produced or consumed at either ends of those shipping lanes.

_____
_____
_____
_____
_____
_____
_____

Instructions for Exploration 9.1 Ocean Waters Short Answer B:

1. Open the **Tides** folder.

2. The *Bermuda* and *Cape Hatteras* placemarks in this folder represent the locations of National Oceanic and Atmospheric Administration tide prediction stations. Click on one of the placemarks then select the *Monthly Tide* prediction link. This will display a chart for the current month, which includes high- and low-tide values for each day measured in the difference from the mean low-water level.

3. Examine the data and determine the date when the range between the highest tide and lowest tide is the greatest.

4. Repeat this evaluation for the other tide prediction station.

Exploration 9.1 Ocean Waters Short Answer B

B. On what date does the highest range at the *Bermuda* placemark occur? What is the range value? On what date does the highest range at the *Cape Hatteras* placemark occur? What is the range value? Is a relationship between the date(s) with the largest range and the current lunar calendar evident? (Search "Moon Phases Calendar" on Google.) How might the location of each tide prediction station help explain any difference in the size of the tidal range between them?

_____

_____

_____

_____

_____

_____

_____

_____

When you complete this exploration, turn off and collapse the **9.1 Ocean Waters** folder.

Exploration 9.2 – Permanent Ice
Multiple Choice

The world's cryosphere, or permanent ice, represents a small amount of the global water budget, but it is exceptionally important as both a regulator of global temperature and as a resource for human populations. The cryosphere includes components of ice on land and ice floating in the ocean.

Instructions for all Parts:

1. Make sure you have opened the **KMZ** file from www.mygeoscienceplace.com.

2. From the Places panel, expand **9. The Hydrosphere.kmz** and then open the **9.2 Permanent Ice** folder.

Instructions for Exploration 9.2 Permanent Ice Part A:

1. Turn on and double-click the **Oceanic Ice** folder.

2. Examine each of the five placemarks labeled *A* through *E* and consider the status of ice at that location. Utilize the imagery at the scale presented.

Exploration 9.2 Permanent Ice Part A

A. Which of the locations in the **Oceanic Ice** folder would be considered an ice shelf?

1. A
2. B
3. C
4. D
5. E

When you complete this part, turn off and collapse the **Oceanic Ice** folder.

Instructions for Exploration 9.2 Permanent Ice Part B:

1. Turn on and double-click the **Greenland Surface Melt** folder. This folder includes data from the National Snow and Ice Data Center showing the annual surface ice melt on the Greenland ice sheet from 1979 to 2007.

2. Use the time slider to view data from each year of data on Greenland's ice sheet surface melt. As you view the data, be sure to compare the spatial patterns with the Maximum Melt Extent chart displayed in the lower left of the 3D Viewer.

3. Evaluate the patterns of surface melt on Greenland's ice sheet from 1979 through 2007, paying particular attention to annual fluctuations in melt days.

Exploration 9.2 Permanent Ice Part B

B. Which of the following statement is best supported by the evidence in the **Greenland Surface Melt** folder?

1. In the period 1979—2007, the greatest amount of melting occurred in 1992.
2. The greatest amount of annual melting typically takes place along Greenland's eastern flank.
3. The maximum annual melt extent trended steadily higher from 1979 to 2007.
4. During the period 1979–2007, every successive year produced greater annual melt.
5. Melt periods greater than 60 days are encountered only in areas south of Nuuk.

When you complete this part, turn off and collapse the **Greenland Surface Melt** folder.

Instructions for Exploration 9.2 Permanent Ice Part C:

1. Double-click the **NSIDC** folder. This folder contains data from the National Snow and Ice Data Center for Sea Ice Extent in the Polar Regions during March and September from 1978 through 2011.

2. Select the **September** folder and use the time slider to watch the animation of sea ice extent in both the Arctic and Antarctic polar regions.

3. Repeat the same examination for the data in the **March** folder.

4. Compare the extent of sea ice in both September and March for both polar regions, noting significant trends or differences between seasons.

Exploration 9.2 Permanent Ice Part C

C. Which of the following statements is best supported by the data in the **NSIDC** folder?

1. Antarctic sea ice usually reaches its greatest annual extent in March.
2. Summer melting of the Arctic sea ice since 2000 has been greater closer to the coasts of Siberia and Alaska rather than Greenland.
3. Arctic sea ice usually reaches its greatest annual extent in September.
4. The fall sea ice in Antarctica generally maintains a 160 km buffer around the Antarctic landmass.
5. The monthly median extent of Arctic ice doesn't expand more than 500 km between September and March.

When you complete this part, turn off and collapse the **NSIDC** folder.

Instructions for Exploration 9.2 Permanent Ice Part D:

1. Open the **Mountain Ice** folder.

2. The five placemarks labeled *A* through *E* each show features with similar color and contrast. Errors occur in image interpretation when viewers mistake a landscape such as salt flat or mineral deposit for snow or ice simply because it appears white. Relying on only one variable without considering the context of the landscape can make interpretation in Google Earth™ difficult.

3. Examine each of the five placemarks and consider the status of ice at that location, being sure to note all appropriate landscape clues.

Exploration 9.2 Permanent Ice Part D

D. Which of the five placemarks in the **Mountain Ice** folder represents mountain ice?

1. A
2. B
3. C
4. D
5. E

When you complete this exploration, turn off and collapse the **9.2 Permanent Ice** folder.

Exploration 9.2 – Permanent Ice
Short Answer

Instructions for all Parts:

1. Make sure you have opened the **KMZ** file from www.mygeoscienceplace.com.

2. From the Places panel, expand **9. The Hydrosphere.kmz** and then open the **9.2 Permanent Ice** folder.

Instructions for Exploration 9.2 Permanent Ice Short Answer A:

1. Open the **NSIDC** folder. This folder contains data from the National Snow and Ice Data Center for Sea Ice Extent in the Polar Regions during March and September from 1978 through 2011.

2. Select the **September** folder and use the time slider to watch the animation of sea ice extent in the Arctic polar regions.

3. Repeat the same examination for the data in the **March** folder.

Exploration 9.2 Permanent Ice Short Answer A

A. Based on trends seen in sea ice extent in the Arctic Ocean, make an assessment of the potential sea ice extant 20 years in the future. What economic and geopolitical impacts might this change produce?

_____

_____

_____

_____

_____

_____

_____

Instructions for Exploration 9.2 Permanent Ice Short Answer B:

1. Ensure that the **NSIDC** folder is open. Examine the trends in sea ice extents for the polar regions from 1978 through 2011.

2. Open the **Greenland Surface Melt** folder. Examine this data, which shows the seasonal changes in Greenland's ice sheet from 1979 to 2007.

3. Compare and contrast the trends seen in both the **NSIDC** and **Greenland Surface Melt** folders. Note the rate of any perceived changes and any specific local or regional patterns.

Exploration 9.2 Permanent Ice Short Answer B

B. What impacts will continuation of the trends seen in the **NSIDC** and **Greenland Surface Melt** folders have on overall global sea levels? Be sure to indicate the differences that changes in the cryosphere of the Antarctic, Arctic, and Greenland would produce.

_____

_____

_____

_____

_____

_____

_____

When you complete this exploration, turn off and collapse the **9.2 Permanent Ice** folder.

Exploration 9.3 – Surface Waters
Multiple Choice

Only a small fraction of Earth's hydrosphere is contained in surface water bodies such as lakes, streams, or marshes. However, these sources of water are particularly important to humans as they represent the most accessible elements of the hydrosphere.

Instructions for all Parts:

1. Make sure you have opened the **KMZ** file from www.mygeoscienceplace.com.

2. From the Places panel, expand **9. The Hydrosphere.kmz** and then open the **9.3 Surface Waters** folder.

Instructions for Exploration 9.3 Surface Waters Part A:

1. Open the **Reservoirs** folder.

2. Examine each of the five placemarks labeled *A* through *E*, noting the characteristics of any potential bodies of water in the area.

Exploration 9.3 Surface Waters Part A

A. Which of the placemarked locations in the **Reservoirs** folder is ***not*** a human-made reservoir?

1. A
2. B
3. C
4. D
5. E

When you complete the previous part, turn off and collapse the **Reservoirs** folder.

Instructions for Exploration 9.3 Surface Waters Part B:

1. Open the **Swamps** folder.

2. Examine the five placemarks labeled *A* through *E*, paying close attention to the landscape context at each location.

Exploration 9.3 Surface Waters Part B

B. Which of the five placemarked locations in the **Swamps** folder would be properly classified as a <u>marsh</u> rather than a <u>swamp</u>?

1. A
2. B
3. C
4. D
5. E

When you complete the previous part, turn off and collapse the **Swamps** folder.

Instructions for Exploration 9.3 Surface Waters Part C:

1. Turn on and double-click the **Drainage Basin** folder.

2. Using Google Earth™ and any outside resource such as your textbook, determine the stream's associated major drainage basin.

Exploration 9.3 Surface Waters Part C

C. The stream noted by the *Drainage Basin* placemark is part of which of the following drainage basins?

1. Zambezi
2. Congo
3. Niger
4. Nile
5. Amazon

When you complete this part, turn off the **Drainage Basin** folder.

Instructions for Exploration 9.3 Surface Waters Part D:

1. Turn on and double-click the **Aral Sea** folder. This folder contains a series of satellite images showing the Aral Sea basin over the past few decades.

2. Read the hyperlinked information, then view the five time-series images from 1973 through 2006 by turning on each respective layer, followed by the more recent default Google Earth™ imagery.

3. Evaluate the changing patterns seen in the Aral Sea over the past several decades.

Exploration 9.3 Surface Waters Part D

D. Which of the following statements is best supported by the evidence in the imagery and data in the **Aral Sea** folder?

1. At its greatest extent, the Aral Sea measured 500 km by 400 km.
2. The primary feeder streams for the Aral Sea entered from the south and east.
3. The Aral Sea was once the world's second largest inland sea.
4. An attempt to save the southern parts of the Aral Sea has been made by damming three key locations.
5. The most dramatic reductions in the Aral Sea have been found along the northern and western shorelines.

When you complete this exploration, turn off and collapse the **9.3 Surface Waters** folder.

Exploration 9.3 – Surface Waters
Short Answer

Instructions for Exploration 9.3 Surface Waters Short Answer A:

1. Make sure you have opened the **KMZ** file from www.mygeoscienceplace.com.

2. From the Places panel, expand **9. The Hydrosphere.kmz** and then open the **9.3 Surface Waters** folder.

3. Turn on and double-click the **Aral Sea** folder. This folder contains a series of satellite images showing the Aral Sea basin over the past few decades.

4. Read the hyperlinked information, then view the five time-series images from 1973 through 2006 by turning on each respective layer. Compare these images to the more recent default Google Earth™ imagery.

5. Use an outside source to investigate the local impacts of the shrinking of the Aral Sea.

Exploration 9.3 Surface Waters Short Answer A

A. Based on the available evidence in the **Aral Sea** folder, Google Earth™ default imagery, and any outside sources, how have local populations been impacted by the dramatic change in the Aral Sea from the 1970s to present? Consider economic, health, and environmental issues that may have arisen as the water retreated. Provide evidence of these impacts in the Google Earth™ imagery.

_____

_____

_____

_____

_____

_____

_____

When you complete this part, collapse and turn off the **Aral Sea** folder.

Instructions for Exploration 9.3 Surface Waters Short Answer B:

1. Zoom in to your region in Google Earth™. Search the area for the two largest non-oceanic bodies of water.

2. Examine the evidence in the imagery to determine the status of these water bodies.

Exploration 9.3 Surface Waters Short Answer B

B. Indicate the latitude and longitude of the two largest non-oceanic water bodies in your area. Are these features lakes or reservoirs? Explain the factors that may contribute to the existence of these features in your region. Be sure to indicate your location in your response.

_____

_____

_____

_____

_____

_____

_____

When you complete this exploration, turn off and collapse the **9.3 Surface Waters** folder.

Exploration 9.4 – Underground Water
Multiple Choice

A final component of the hydrosphere is much less conspicuous, but important nonetheless. The presence of underground water can be revealed by human adaptations and displays anthropogenic features.

Instructions for all Parts:

1. Make sure you have opened the **KMZ** file from www.mygeoscienceplace.com.

2. From the Places panel, expand **9. The Hydrosphere.kmz** and then open the **9.4 Underground Water** folder.

Instructions for Exploration 9.4 Underground Water Part A:

1. Open the **Irrigation** folder. The five placemarked locations labeled *A* through *E* show different agricultural landscapes in various parts of the United States and Western Europe.

2. Examine the five placemarked locations, paying particular attention to the potential sources for agricultural water at each site.

Exploration 9.4 Underground Water Part A

A. Which of the placemarked locations in the **Irrigation** folder is the most likely to utilize pumped groundwater for irrigation?

   1. A
   2. B
   3. C
   4. D
   5. E

When you complete this part, turn off and collapse the **Irrigation** folder.

Instructions for Exploration 9.4 Underground Water Part B:

1. Turn on and double-click the **Groundwater Levels** folder. This data shows the status of groundwater at selected monitoring sites across the United States. The color coding indicates the groundwater level compared with normal levels at each site.

2. Turn on the *Ogallala* layer for a spatial comparison of groundwater monitoring sites and the Great Plains aquifer.

3. Assess the regional pattern of groundwater depth measurements at reporting well sites across the country.

Exploration 9.4 Underground Water Part B

B. Which of the following statements is best supported by the evidence presented in the *Groundwater* and *Ogallala* layers?

1. Based on the monitoring sites, groundwater is distributed evenly across the continental United States.
2. The Ogallala aquifer underlies part of ten different states.
3. Groundwater is not monitored in the greater New York metropolitan area.
4. Groundwater monitoring stations are not present in arid regions of the United States.
5. Groundwater is monitored along many streams and rivers such as the Mississippi, Arkansas, and Snake.

When you complete this part, turn off the **Groundwater Levels** folder and *Ogallala* layer.

Instructions for Exploration 9.4 Underground Water Part C:

1. Turn on and double-click the **Aquifer Rock Types** folder. This layer displays a generalized pattern of aquifer rock types.

2. Turn on the *Ogallala* layer and compare its extent with the distribution of aquifer rock types.

Exploration 9.4 Underground Water Part C

C. Based on the data in the **Aquifer Rock Type** folder, what rock type dominates the Ogallala (High Plains) aquifer?

1. Carbonate-rock
2. Unconsolidated sand and gravel
3. Igneous and metamorphic
4. Sandstone
5. Sandstone and carbonate

When you complete this part, turn off and collapse the **Aquifer Rock Type** folder and turn off the *Ogallala* layer.

Instructions for Exploration 9.4 Underground Water Part D:

1. Turn on and double-click the **Great Man-made River** folder. This folder contains data on a massive water diversion project undertaken in Libya over the last few decades. Click the hyperlink on the **Great Man-made River** folder to read a brief description of the project.

2. Examine the imagery layers for *Site 1* and *Site 2*, locations associated with the project. Be certain to consider the evidence of changes at both sites as seen in the imagery.

Exploration 9.4 Underground Water Part D

D. Which of the following statements is best supported by the images and data in the **Great Man-made River** folder?

1. The Great Man-made River obtains its water from a diversion of the Nile River.
2. The Great Man-made River has diverted water away from and subsequently decreased agricultural activities at locations such as Al-Jawf.
3. The Great Man-made River has led to the construction of two large reservoirs northwest of Suluq.
4. Prior to the construction of the Great Man-made River, Libya had no agriculture.
5. At current extraction rates, the Nubian sandstone aquifer system will be depleted in less than 30 years.

When you complete this exploration, turn off and collapse the **9.4 Underground Water** folder.

Exploration 9.4 – Underground Water
Short Answer

Instructions for all Parts:

1. Make sure you have opened the **KMZ** file from www.mygeoscienceplace.com.

2. From the Places panel, expand **9. The Hydrosphere.kmz** and then open the **9.4 Underground Water** folder.

Instructions for Exploration 9.4 Underground Water Short Answer A:

1. Turn on the **Groundwater Levels** folder.

2. Use the Search panel to locate and zoom to Seward County, Kansas.

3. Click on a well site from within Seward County. In the pop-up that opens, click the well number hyperlink to access detailed data from this site.

4. Review the Periodic Groundwater data section, noting any trends in the water level at this well site.

5. Repeat the same examination for two additional well sites in Seward County.

Exploration 9.4 Underground Water Short Answer A

A. Record the site number of your three selected wells and describe the trend, if any, seen in the periodic groundwater data from each.

_____

_____

_____

_____

_____

_____

_____

Instructions for Exploration 9.4 Underground Water Short Answer B:

1. Use the Search panel to zoom to a view that contains both Sheridan County and Box Butte County in northwest Nebraska.

2. Examine the landscape of this region, paying particular attention to the patterns of agriculture and the location of groundwater monitoring wells.

Exploration 9.4 Underground Water Short Answer B

B. Describe the locations of the groundwater monitoring sites in these two counties and provide an explanation for the lack of monitoring sites in most of southern Sheridan County.

_____

_____

_____

_____

_____

_____

_____

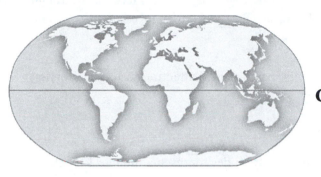

Encounter Physical Geography

Name:_____

Date: _____

Chapter 10
Cycles and Patterns in the Biosphere

Exploration 10.1 – Biogeochemical Cycles
Multiple Choice

The world's biological diversity reflects the flow of energy, water, and nutrients through a series of biogeochemical cycles. These cycles vary in different parts of the world at different times of the year as a result of latitudinal variations and local geographic characteristics.

Instructions for all Parts:

1. Make sure you have opened the **KMZ** file from www.mygeoscienceplace.com.

2. From the Places panel, expand **10. The Hydrosphere.kmz** and then open the **10.1 Biogeochemical Cycles** folder.

Instructions for Exploration 10.1 Biogeochemical Cycles Part A:

1. Turn on the **Insolation** layer. This layer displays the rates of insolation in watts per square meter in April 2012.

2. Examine the pattern of insolation, noting any significant local or regional differences.

Exploration 10.1 Biogeochemical Cycles Part A

A. Which of the following statements is best supported by the data in the *Insolation* layer?

1. The cloud cover of the ITCZ limits insolation just south of the equator at this time.
2. The regions of highest and lowest insolation in the contiguous United States are both located in the western part of the country.
3. At this time of year, insolation is generally greater in the Antarctic polar region than the Arctic polar region.
4. North Africa receives substantially less insolation than Sub-Saharan Africa in April 2012.
5. The slopes of the Himalayas represent an area of uniform insolation.

When you complete this part, turn off the **Insolation** layer.

Instructions for Exploration 10.1 Biogeochemical Cycles Part B:

1. Open the **Primary Productivity** folder. Each of the five image layers labeled *A* through *E* represents the average primary vegetative productivity in grams of carbon per square meter for a different month of the calendar year.

2. Examine the patterns in the five image layers, paying close attention to the distribution and intensity of primary productivity in each.

Exploration 10.1 Biogeochemical Cycles Part B

B. Which of the image layers in the **Primary Productivity** folder represents the latest month in the calendar year?

1. A
2. B
3. C
4. D
5. E

When you complete this part, turn off and collapse the **Primary Productivity** folder.

Instructions for Exploration 10.1 Biogeochemical Cycles Part C:

1. Open the **NDVI and Chlorophyll** folder.

2. The *Chlorophyll* layer displays recorded chlorophyll readings in milligrams per square meter for April 2012; the *NDVI* layer displays the Normalized Difference Vegetation Index scale of vegetative growth for April 2012. Both layers are measures of biological production. Alternate between the layers to compare the data in each.

3. Examine the patterns seen in the *NDVI* and *Chlorophyll* layers noting any significant regional patterns and variations.

Exploration 10.1 Biogeochemical Cycles Part C

C. Which of the following statements is best supported by the data in the **NDVI and Chlorophyll** folder?

1. The highest levels of chlorophyll are typically found in the open ocean rather than along coastlines.
2. The Mediterranean Sea has relatively higher chlorophyll levels compared to the Baltic Sea.
3. At the time of this image, Sub-Saharan Africa has lower vegetative growth than North Africa.
4. The Democratic Republic of the Congo generally has the lowest levels of vegetative growth in Africa at this time of year.
5. In this dataset, Lake Erie has relatively higher levels of chlorophyll than the rest of the Great Lakes.

When you complete this part, turn off and collapse the **NDVI and Chlorophyll** folder.

Instructions for Exploration 10.1 Biogeochemical Cycles Part D:

1. Open the **Leaf Index** folder. The two layers in this folder show the total leaf area compared to total ground area in square meter divided by square meter for two seasons.

2. Alternate between the *Leaf Index 1* and *Leaf Index 2* layers to view the patterns of vegetation, noting that dark-green areas represent a relatively high leaf area while brown areas are relatively low. Consider what the different patterns may indicate about the time of year for each layer.

Exploration 10.1 Biogeochemical Cycles Part D

D. Which of the following pairings of *Leaf Index* layer to month is the most likely based on the data shown by the layers in the **Leaf Index** folder?

1. Leaf Index 1 = January, Leaf Index 2 = July
2. Leaf Index 1 = July, Leaf Index 2 = January
3. Leaf Index 1 = January, Leaf Index 2 = February
4. Leaf Index 1 = June, Leaf Index 2 = July
5. Leaf Index 1 = March, Leaf Index 2 = September

When you complete this exploration, turn off and collapse the **10.1 Biogeochemical Cycles** folder.

Exploration 10.1 – Biogeochemical Cycles
Short Answer

Instructions for all Parts:

1. Make sure you have opened the **KMZ** file from www.mygeoscienceplace.com.

2. From the Places panel, expand **10. The Hydrosphere.kmz** and then open the **10.1 Biogeochemical Cycles** folder.

Instructions for Exploration 10.1 Biogeochemical Cycles Short Answer A:

1. Turn on and double-click the *Energy Flow* Gigapan image to fly into the scene.

2. Examine the living and non-living elements of this landscape, paying attention to examples of energy flow mechanisms present in the scene.

Exploration 10.1 Biogeochemical Cycles Short Answer A

A. In a short paragraph, describe the potential energy flow mechanisms that are present in the scene shown in the *Energy Flow* Gigapan image.

_____

_____

_____

_____

_____

_____

_____

Instructions for Exploration 10.1 Biogeochemical Cycles Short Answer B:

1. Turn on and double-click the *Carbon Cycle* Gigapan image to fly into the scene.

2. Examine the elements of this landscape, noting any potential paths of the carbon cycle that may be present.

Exploration 10.1 Biogeochemical Cycles Short Answer B

B. In a short paragraph, describe the various paths of the carbon cycle that could be utilized by elements of the natural and anthropogenic landscape seen in the *Carbon Cycle* Gigapan image.

_____

_____

_____

_____

_____

_____

_____

When you complete this exploration, turn off and collapse the **10.1 Biogeochemical Cycles** folder.

Exploration 10.2 – Environmental Relationships I
Multiple Choice

Environmental relationships occur at a variety of scales from microscopic to global. These relationships help explain the distribution of biotic phenomena such as plants and animals.

Instructions for all Parts:

1. Make sure you have opened the **KMZ** file from www.mygeoscienceplace.com.

2. From the Places panel, expand **10. The Hydrosphere.kmz** and then open the **10.2 Environmental Relationships I** folder.

Instructions for Exploration 10.2 Environmental Relationships I Parts A–B:

1. Turn on and double-click the **Antarctica Food Web** folder. Each of the placemarks in this folder represents a component of the food web in the Antarctic polar region.

2. Explore this regional food web, beginning by clicking on the *Phytoplankton* link. Examine the characteristics of each animal, noting their relative position in the Antarctic food web.

Exploration 10.2 Environmental Relationships I Part A

A. Which of the following animals indicated in the **Antarctica Food Web** folder utilizes the Crabeater Seal as a source of food?

1. Krill
2. Squid
3. Sharks
4. Leopard Seal
5. Humpback Whale

Exploration 10.2 Environmental Relationships I Part B

B. Which of the following animals indicated in the **Antarctica Food Web** folder is a source of food for the greatest number of other animals in this food web?

1. Phytoplankton
2. Krill
3. Fish
4. Squid
5. Killer Whales

When you complete the previous two parts, collapse and turn off the **Antarctica Food Web** folder.

Instructions for Exploration 10.2 Environmental Relationships I Part C:

1. Open the **Light** folder.

2. Double-click on each of the five placemarks labeled *A* through *E* one at a time to evaluate the relative location of each place. Be sure to note specific site characteristics such as latitude and longitude.

Exploration 10.2 Environmental Relationships I Part C

C. Which of the five placemarked locations in the **Light** folder would have the most consistent light on a year-round basis?

1. A
2. B
3. C
4. D
5. E

When you complete this part, collapse and turn off the **Light** folder.

Instructions for Exploration 10.2 Environmental Relationships I Part D:

1. Open the **Temperature** folder.

2. Double-click on each of the five placemarks labeled *A* through *E* one at a time to evaluate the relative location of each place. Be sure to note specific site characteristics such as latitude and elevation.

Exploration 10.2 Environmental Relationships I Part D

D. Based on the factors of latitude and elevation, which of the five locations in the **Temperature** folder would provide the most consistently warm temperatures throughout the year?

1. A
2. B
3. C
4. D
5. E

When you complete this exploration, turn off and collapse the **10.2 Environmental Relationships I** folder.

Exploration 10.2 – Environmental Relationships I
Short Answer

Instructions for all Parts:

1. Make sure you have opened the **KMZ** file from www.mygeoscienceplace.com.

2. From the Places panel, expand **10. The Hydrosphere.kmz** and then open the **10.2 Environmental Relationships I** folder.

Instructions for Exploration 10.2 Environmental Relationships I Short Answer A:

1. Double-click the *Environmental Constraints* placemark.

2. Examine the scene displayed at the placemark, being sure to consider any factor that might impact the distribution of local flora.

Exploration 10.2 Environmental Relationships I Short Answer A

A. Write a short explanation for the distribution of flora seen at the *Environmental Control* placemark. Be sure to consider factors such as light, moisture, temperature, and wind in your response.

_____

_____

_____

_____

_____

_____

_____

Instructions for Exploration 10.2 Environmental Relationships I Short Answer B:

1. Use Google Earth™ to locate an area of natural vegetation near your location.

2. Examine that landscape, making note of any factors that might impact the patterns of flora and fauna.

Exploration 10.2 Environmental Relationships I Short Answer B

B. Describe some of the environmental constraints on biota that may occur at your location. Be sure to provide the coordinates at the location you've chosen.

_____

_____

_____

_____

_____

When you complete this exploration, turn off and collapse the **10.2 Environmental Relationships I** folder.

Exploration 10.3 – Environmental Relationships II
Multiple Choice

Environmental relationships manifest in the physical world through processes such as plant succession and invasive species. These phenomena speak to the ongoing competition occurring between different species as they seek energy, light, and nutrients.

Instructions for all Parts:

1. Make sure you have opened the **KMZ** file from www.mygeoscienceplace.com.

2. From the Places panel, expand **10. The Hydrosphere.kmz** and then open the **10.3 Environmental Relationships II** folder.

Instructions for Exploration 10.3 Environmental Relationships II Part A:

1. Turn on and double-click the **Succession** folder. The five placemarks labeled *A* through *E* indicate locations that have undergone the dynamic process of plant succession near an abandoned river meander.

2. Examine the five placemarked locations, paying attention to the vegetation at each site.

Exploration 10.3 Environmental Relationships II Part A

A. Which of the placemarked locations in the **Succession** folder would likely be the farthest along in the plant succession process as the former river channel dries up?

1. A
2. B
3. C
4. D
5. E

When you complete this part, turn off and collapse the **Succession** folder.

Instructions for Exploration 10.3 Environmental Relationships II Parts B–C:

1. From the *Primary Database*, open the *Gallery* layer then open the *NASA* layer. Open the *Earth City Lights* layer then turn on and double-click the *Earth City Lights* layer. Nighttime illumination is significant because exposure to non-natural light sources has physiological implications for plants and animals.

2. Examine the distribution of nighttime illumination around the world, noting concentrations of light and darkness.

Exploration 10.3 Environmental Relationships II Part B

B. Which of the following statements is best supported by the evidence in the *Earth City Lights* layer?

1. The western half of the United States has more light pollution than the eastern half of the United States.
2. The largest cluster of lights in the United States occurs along the northern Gulf Coast.
3. Biota in North Korea is less likely to be impacted by light pollution than biota in South Korea.
4. South Asia is the least illuminated region.
5. Light pollution is evenly distributed across Egypt.

Exploration 10.3 Environmental Relationships II Part C

C. Which of the following statements is **not** supported by the evidence in the *Earth City Lights* layer?

1. Lights are limited in the Everglades region of Florida.
2. The strings of lights leading into the city of São Paulo, Brazil, follow major river channels.
3. Lights are clustered along the Nile River in Egypt.
4. The general lack of light in the northern parts of Mali can be partially explained by the presence of the Sahara Desert.
5. The lack of lights in the majority of Bhutan can be explained by the presence of mountainous terrain throughout that country.

When you complete the previous two parts, turn off the *Earth City Lights* layer.

Instructions for Exploration 10.3 Environmental Relationships II Part D:

1. Turn on and double-click the **Invasive Species** folder. The green states represent states where the kudzu vine has been reported.

2. Turn on and double-click the *Kudzu* placemark to fly into a street-view perspective of a location in Georgia with a large amount of kudzu.

3. Evaluate the map of kudzu distribution and the status of kudzu from the street-view perspective.

Exploration 10.3 Environmental Relationships II Part D

D. Based on the evidence found in the **Invasive Species** folder, which of the following climate characteristics best match the prime growing environment of kudzu?

1. warm & humid
2. cold & humid
3. warm & dry
4. cold & dry
5. hot & dry

When you complete this exploration, turn off and collapse the **10.3 Environmental Relationships II** folder.

Exploration 10.3 – Environmental Relationships II
Short Answer

Instructions for Exploration 10.3 Environmental Relationships II Short Answer A:

1. Make sure you have opened the **KMZ** file from www.mygeoscienceplace.com.

2. From the Places panel, expand **10. The Hydrosphere.kmz** and then open the **10.3 Environmental Relationships II** folder.

3. Turn on and double-click the **Mount Saint Helens** folder. The three images contained in this folder were captured by the Landsat satellite and show the Mount Saint Helens region before, immediately following, and 15 years after the volcano's eruption in 1980.

4. Examine the three Landsat images from 1972, 1983, and 1999, as well as current and historic imagery available in Google Earth™.

Exploration 10.3 Environmental Relationships II Short Answer A

A. Identify a location in the **Mount Saint Helens** folder imagery using latitude and longitude and describe the pattern of succession that has been experienced in that region.

_____

_____

_____

_____

_____

_____

_____

When you complete this part, turn off and collapse the **Mount Saint Helens** folder.

Instructions for Exploration 10.3 Environmental Relationships II Short Answer B

1. From the *Primary Database*, open the *Gallery* layer then open the *NASA* layer. Open the *Earth City Lights* layer then turn on and double-click the *Earth City Lights* layer. Nighttime illumination is significant because exposure to non-natural light sources has physiological implications for plants and animals.

2. Examine the distribution of nighttime illumination around the world, noting concentrations of light and darkness.

3. Select a location and analyze the distribution of lights in comparison to physical features.

Exploration 10.3 Environmental Relationships II Short Answer B

B. Identify the location you have examined by latitude and longitude, and describe the patterns of light and physical terrain.

_____

_____

_____

_____

_____

_____

_____

When you complete this exploration, turn off and collapse the **10.3 Environmental Relationships II** folder.

Exploration 10.4 – Wildfire
Multiple Choice

Most environmental relationships factors, such as those related to the influence of light, moisture, or temperature, occur as incremental background processes. However, some factors occur during sudden, severe events. Wildfire is an example of such a process. Evidence of wildfire events in data and imagery and impacts of fire on biologic processes provide us with clues to the scope and extent of this phenomenon.

Instructions for all Parts:

1. Make sure you have opened the **KMZ** file from www.mygeoscienceplace.com.

2. From the Places panel, expand **10. The Hydrosphere.kmz** and then open the **10.4 Wildfire** folder.

Instructions for Exploration 10.4 Wildfire Part A:

1. Turn on and open the **Historic Fires** folder. The five placemarked locations in this folder represent sites where wildfires may have occurred since the beginning of the 21st century.

2. Examine the five placemarked locations, making sure to utilize the historical imagery to identify any evidence of wildfires at each location

Exploration 10.4 Wildfire Part A

A. Based on the evidence in the historical imagery, which of the placemarks in the **Historic Fires** folder does **not** show evidence of a significant wildfire event?

1. Hayman
2. Kinglake
3. Les Cheneaux
4. Mount Carmel
5. Rodeo

When you complete this part, turn off and collapse the **Historic Fire** folder.

Instructions for Exploration 10.4 Wildfire Part B:

1. Turn on and double-click the **Yellowstone** folder. The five placemarks labeled *A* through *E* are located at various points throughout the Yellowstone National Park region.

2. Utilizing the available imagery for the scale shown, evaluate each placemarked location, noting any evidence of significant fire events.

Exploration 10.4 Wildfire Part B

B. Based on the available evidence at the scale shown, which placemarked location in the **Yellowstone** folder has the best evidence of a significant fire event?

1. A
2. B
3. C
4. D
5. E

When you complete this part, turn off and collapse the **Yellowstone** folder.

Instructions for Exploration 10.4 Wildfire Part C:

1. Turn on and double-click the *Perth* image overlay. This satellite image shows the region around the Western Australian city of Perth during a period of active wildfires.

2. Zoom in to the area immediately around Perth and evaluate the status of wildfires, being sure to note the likely direction of the movement of the fires.

Exploration 10.4 Wildfire Part C

C. Based on the evidence seen in the *Perth* image overlay, in which direction are the fires around the city of Perth moving?

1. North
2. South
3. East
4. West
5. The direction cannot be determined from the image.

When you complete this part, turn off the *Perth* image overlay.

Instructions for Exploration 10.4 Wildfire Part D:

1. Turn on and double-click the **Fire Seasonality** folder. The three layers in this folder show active fires detected by MODIS satellite sensors for three dates: December 27, 2011, March 27, 2012, and June 12, 2012.

2. Toggle between each of the dates, examining the difference in fire distribution patterns and noting any significant seasonal trends.

Exploration 10.4 Wildfire Part D

D. Which of the following statements is best supported by the data shown in the **Fire Seasonality** folder?

1. Wildfire events were much more frequent in December 2011 than in March 2012.
2. Wildfire events had the greatest geographical extent in December 2011.
3. The highest number of wildfire events occurred in June 2012.
4. The peak of the 2012 fire season on the Great Plains occurred in early spring.
5. The highest concentration of wildfire events in Mexico during June 2012 was found in the Sierra Madre Oriental.

When you complete this exploration, turn off and collapse the **10.4 Wildfire** folder.

Exploration 10.4 – Wildfire
Short Answer

Instructions for all Parts:

1. Make sure you have opened the **KMZ** file from www.mygeoscienceplace.com.

2. From the Places panel, expand **10. The Hydrosphere.kmz** and then open the **10.4 Wildfire** folder.

3. Open the **Wildfire Forecast** folder then turn on the **Significant Fire Potential** folder. This folder links directly to the United States Forest Service's Active Fire Mapping Program and will take a few moments to load. Once the information appears, you should find several folders inside that include information about wildfire potential over the next seven-day period.

4. Open the **Significant Fire Potential** folder, turn on the **Legends and Logos** folder, and then examine the **Forecast** folders for the available dates. Wildfire potential forecasts are displayed by Predictive Service Areas (PSAs) for the regions of the country actively monitored by this system.

5. Evaluate the data for significant wildfire potential using the data in the **Significant Fire Potential** folder. Click on the PSAs to open informational pop-up windows containing significant wildfire forecast criteria.

Exploration 10.4 Wildfire Short Answer A

A. Describe the pattern of significant wildfire risk across the country for the current date. Indicate the region most at risk. Provide the status of the Fuels Forecast and Ignition Trigger for one of the PSAs in the region at risk. Note the name of the selected PSA.

_____

_____

_____

_____

_____

_____

_____

Exploration 10.4 Wildfire Short Answer B

B. Compare the forecast for the current date and the most distant date. Indicate any significant changes between the current fire potential and the forecast for a week into the future. Provide a possible explanation for the changes, being sure to keep in mind potential changes in precipitation, temperature, and wind direction.

_____

_____

_____

_____

_____

_____

_____

When you complete this exploration, turn off and collapse the **10.4 Wildfire** folder.

Encounter Physical Geography

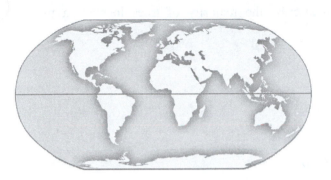

Name:_____

Date: _____

Chapter 11
Terrestrial Flora and Fauna

Exploration 11.1 – Flora
Multiple Choice

The flora, or plants, of the Earth exhibit remarkable diversity across the globe. This diversity is largely a product of adaptations to different environments. Some plants have developed characteristics that help them succeed in particularly dry climates, while others thrive in relatively moist conditions. This exploration samples some of the adaptations flora have made to sustain life in locations around the planet.

Instructions for all Parts:

1. Make sure you have opened the **KMZ** file from www.mygeoscienceplace.com.

2. From the Places panel, expand **11. Terrestrial Flora and Fauna.kmz** and then open the **11.1 Flora** folder.

Instructions for Exploration 11.1 Flora Part A:

1. Turn on and double-click the *Lassen National Park* Gigapan image.

2. Evaluate the scene shown in the image, noting the predominant tree cover at this location.

Exploration 11.1 Flora Part A

A. Which of the following terms would best describe the trees located in the *Lassen National Park* Gigapan Image?

1. deciduous trees
2. hardwood trees
3. evergreen trees
4. broadleaf trees
5. herbaceous trees

Instructions for Exploration 11.1 Flora Part B:

1. Turn on and double-click the *Hawksbill Gap* Gigapan image.
2. Evaluate the scene shown in the image, noting the predominant tree cover at this location.

Exploration 11.1 Flora Part B

B. Which of the following terms would best describe the majority of trees located in the *Hawksbill Gap* Gigapan Image?

1. deciduous trees
2. evergreen trees
3. needleleaf trees
4. softwood trees
5. gymnosperm trees

Instructions for Exploration 11.1 Flora Parts C–D

1. Open the **Environmental Adaptations** folder. The five Gigapan/Gigapixel images in this folder represent a selection of different natural landscapes.

2. Turn on and double-click each of the five images individually. Examine the characteristics of flora in each environment, paying particular attention to adaptations to extreme moisture conditions.

Exploration 11.1 Flora Part C

C. Of the sites included in the **Environmental Adaptations** folder, which one is dominated by plants with xerophytic adaptations?

1. Big Cypress National Preserve
2. Catalina Mountains
3. Devil's Tower
4. Eastern Townships
5. Tarpon Springs

Exploration 11.1 Flora Part D

D. Of the sites included in the **Environmental Adaptations** folder, which one is dominated by plants with hydrophytic adaptations?

1. Big Cypress National Preserve
2. Catalina Mountains
3. Devil's Tower
4. Eastern Townships
5. Tarpon Springs

When you complete this exploration, turn off and collapse the **11.1 Flora** folder.

Exploration 11.1 – Flora
Short Answer

Instructions for all Parts:

1. Make sure you have opened the **KMZ** file from www.mygeoscienceplace.com.

2. From the Places panel, expand **11. Terrestrial Flora and Fauna.kmz** and then open the **11.1 Flora** folder.

Instructions for Exploration 11.1 Flora Short Answer A:

1. Turn on and double-click the *Lassen National Park* Gigapan image. Examine the landscape of this location, noting the predominant type of tree in this area.

2. Repeat the same examination using the *Hawksbill Gap* Gigapan image.

Exploration 11.1 Flora Short Answer A

A. Hypothesize as to why the difference between the type of trees present in the *Lassen National Park* and *Hawksbill Gap* Gigapan images is so stark. What environmental factors could be responsible for the difference?

_____
_____
_____
_____
_____
_____
_____

Instructions for Exploration 11.1 Flora Short Answer B:

1. Open the **Environmental Adaptations** folder.

2. Select one of the Gigapan/Gigapixel images and turn on and double-click it to fly into the scene. Evaluate the landscape shown in the image.

Exploration 11.1 Flora Short Answer B

B. Based on the vegetation shown in the Gigapan image you selected, apply floristic terminology to a description of the landscape as appropriate.

_____

_____

_____

_____

_____

_____

_____

When you complete this exploration, turn off and collapse the **11.1 Flora** folder.

Exploration 11.2 – Spatial Associations of Plants
Multiple Choice

As a response to varied environmental conditions, certain groups of plants are more likely to be found in association with one another. As a result, geographers can map the floristic associations based on the presence of dominant plants. Examples include forests, grasslands, and wetlands.

Instructions for all Parts:

1. Make sure you have opened the **KMZ** file from www.mygeoscienceplace.com.

2. From the Places panel, expand **11. Terrestrial Flora and Fauna.kmz** and then open the **11.2 Spatial Associations of Plants** folder.

Instructions for Exploration 11.2 Spatial Associations of Plants Parts A–B:

1. Turn on and open the **Floristic Associations** folder.

2. Double-click on each of the five placemarked locations labeled *A* through *E* and examine the landscape at each location, noting the predominant floristic association.

Exploration 11.2 Spatial Associations of Plants Part A

A. Which of the placemarks in the **Floristic Associations** folder is the best example of a woodland plant association?

1. A
2. B
3. C
4. D
5. E

Exploration 11.2 Spatial Associations of Plants Part B

B. Which of the placemarks in the **Floristic Associations** folder is the best example of a shrubland plant association?

1. A
2. B
3. C
4. D
5. E

When you complete the previous two parts, turn off and collapse the **Floristic Associations** folder.

Instructions for Exploration 11.2 Spatial Associations of Plants Part C:

1. Turn on and double-click the *Adret-Ubac* placemark.

2. Evaluate the vegetation conditions that exist at this location, noting any differences between opposing sides of valleys and ridges.

Exploration 11.2 Spatial Associations of Plants Part C

C. Which of the following statements is best supported by the evidence seen at the
Adret-Ubac placemark?

1. The southwest-facing ubac slope receives the sun's rays at a lower angle, reducing the amount of energy available to vegetation.
2. Trees on the southwest slope have been removed to make room for new housing developments.
3. The northeast-facing adret slope receives the sun's rays at a higher angle, providing vegetation with more energy.
4. Trees on the northeast slope have been removed to make room for new housing developments.
5. The northeast-facing ubac slope's lower sun angle provides a more suitable growing environment for local trees than the high sun angle of the southwest-facing slope.

When you complete this part, turn off the *Adret-Ubac* placemark.

Instructions for Exploration 11.2 Spatial Associations of Plants Part D:

1. Open the **Riparian Vegetation** folder.

2. Evaluate the five placemarked locations individually, noting the presence of riparian vegetation at each.

Exploration 11.2 Spatial Associations of Plants Part D

D. Which of the placemarked locations in the **Riparian Vegetation** folder is the best example of riparian vegetation?

1. A
2. B
3. C
4. D
5. E

When you complete this exploration, turn off and collapse the **11.2 Spatial Associations of Plants** folder.

Exploration 11.2 – Spatial Associations of Plants
Short Answer

Instructions for all Parts:

1. Make sure you have opened the **KMZ** file from www.mygeoscienceplace.com.

2. From the Places panel, expand **11. Terrestrial Flora and Fauna.kmz** and then open the **11.2 Spatial Associations of Plants** folder.

Instructions for Exploration 11.2 Spatial Associations of Plants Short Answer A:

1. Open the **Vertical Zonation** folder.

2. Double-click on each of the five placemarks labeled *A* through *E* individually, noting the elevation at each location, as listed in the Status Bar. Examine the landscapes at each location and note the significant vegetation patterns present.

Exploration 11.2 Spatial Associations of Plants Short Answer A

A. Document the altitude at each placemarked location in the **Vertical Zonation** folder and describe the vegetation patterns at each. Explain why altitudinal zonation is a factor in the patterns seen.

_____

_____

_____

_____

_____

_____

_____

When you complete this part, turn off and collapse the **Vertical Zonation** folder.

Instructions for Exploration 11.2 Spatial Associations of Plants Short Answer B:

1. Navigate to your location in Google Earth™.

2. Explore your immediate area to locate an example of riparian vegetation. Remember that riparian vegetation can be found along the banks of streams, around lakes, or near other areas of surface water.

Exploration 11.2 Spatial Associations of Plants Short Answer B

B. Provide the latitude and longitude of a site near your location that represents an area of riparian vegetation. Describe the extent of the vegetation cover. How aware were you of this location prior to finding it in Google Earth™? Is it considered a significant environmental resource in your community? Be sure to note your location in your response.

_____

_____

_____

_____

_____

_____

_____

When you complete this exploration, turn off and collapse the **11.2 Spatial Associations of Plants** folder.

Exploration 11.3 – Spatial Patterns of Fauna
Multiple Choice

Terrestrial fauna is rarely studied by geographers to the degree of terrestrial flora. Animals are less prominent, fewer in number, and mobile. These factors discourage the use of fauna for generalized assessment of biotic regions. Nonetheless, the animals of the Earth exhibit environmental adaptations to enable their survival in varied niches.

Instructions for all Parts:

1. Make sure you have opened the **KMZ** file from www.mygeoscienceplace.com.

2. From the Places panel, expand **11. Terrestrial Flora and Fauna.kmz** and then open the **11.3 Spatial Patterns of Fauna** folder.

Instructions for Exploration 11.3 Spatial Patterns of Fauna Parts A–C:

1. Open the **Animals** folder. The five placemarked locations labeled *A* through *E* show different landscapes that include the presence of animals.

2. Examine each of the placemarks, paying close attention to the type of animal found at each location. Be sure to consider clues such as physical location, climate characteristics, local vegetation, and zoogeographic region.

Exploration 11.3 Spatial Patterns of Fauna Part A

A. At which of the placemarked locations in the **Animals** folder do you find camels?

    1. A
    2. B
    3. C
    4. D
    5. E

Exploration 11.3 Spatial Patterns of Fauna Part B

B. At which of the placemarked locations in the **Animals** folder do you find seals?

    1. A
    2. B
    3. C
    4. D
    5. E

Exploration 11.3 Spatial Patterns of Fauna Part C

C. Which of the following zoogeographic regions has the greatest representation amongst the sites placemarked in the **Animals** folder?

    1. Ethiopian
    2. Nearctic
    3. Neotropical
    4. Oriental
    5. Madagascar

When you complete the previous three parts, turn off and collapse the **Animals** folder.

Instructions for Exploration 11.3 Spatial Patterns of Fauna Part D:

1. Double-click the **Kafue Flats** folder. This folder contains data on a region in central Zambia along the Kafue River.

2. Click the *Kafue Wetlands* hyperlink to read about the changes that have taken place in this area and the impacts on the local fauna.

3. Alternate between the *1973* and *2007* image overlays to compare the region before and after these changes. Turn off both overlays to examine the region with the default Google Earth™ imagery.

Exploration 11.3 Spatial Patterns of Fauna Part D

D. Which of the following statements is best supported by the imagery in the **Kafue Flats** folder?

1. Four major dams have been constructed on this section of the Kafue River since 1973.
2. Endemic species such as the Kafue antelope and Wattled crane live in the impacted area.
3. The area upstream from the Itezhi-Tezhi Dam is where the most significant seasonal flooding occurs.
4. The length of the Kafue River that experiences seasonal flooding cannot exceed 100 kilometers due to the presence of dams.
5. The Kafue Gorge Dam's impoundment has a much larger area than the Itezhi-Tezhi Dam's impoundment.

When you complete this exploration, turn off and collapse the **11.3 Spatial Patterns of Fauna** folder.

Exploration 11.3 Spatial Patterns of Fauna
Short Answer

Instructions for Exploration 11.3 Spatial Patterns of Fauna Short Answer A:

1. From the *Primary Database*, expand the *Global Awareness* layer. Turn on the *ARKive: Endangered Species* layer.

2. Numerous ARKive logos will appear on the globe in the 3D Viewer at locations where specific animals are threatened.

3. Select a location and click on the *ARKive* placemark to view information about a threatened species.

Exploration 11.3 Spatial Patterns of Fauna Short Answer A

A. Describe an animal from the *ARKive: Endangered Species* layer. Include its primary zoogeographic region, endangered status, and principal threats.

_____

_____

_____

_____

_____

_____

When you complete this part, turn off the *ARKive: Endangered Species* layer from the *Primary Database*.

Instructions for Exploration 11.3 Spatial Patterns of Fauna Short Answer B:

1. Make sure you have opened the **KMZ** file from www.mygeoscienceplace.com.

2. From the Places panel, expand **11. Terrestrial Flora and Fauna.kmz** and then open the **11.3 Spatial Patterns of Fauna** folder.

3. Open the **Animals** folder. Select two of the placemarks and examine the animals present in each, noting the significant similarities or differences in environmental factors and adaptations by the animals.

Exploration 11.3 Spatial Patterns of Fauna Short Answer B

B. Describe the similarities and differences in habitat between two placemarks in the *Animals* folder. Be sure to indicate any specific environmental adaptations that the species present at each placemark may utilize to survive in those locations. Consult outside resources if necessary.

_____

_____

_____

_____

_____

_____

_____

When you complete this exploration, turn off and collapse the **11.3 Spatial Patterns of Fauna** folder.

Exploration 11.4 – Biomes
Multiple Choice

Encompassing both flora and fauna, along with abiotic constituents such as soil, climate, and topography, biomes can be considered regional-scale ecosystems. While natural distributions of plants and animals predominantly define biomes, those distributions are actively modified by human activity.

Instructions for all Parts:

1. Make sure you have opened the **KMZ** file from www.mygeoscienceplace.com.

2. From the Places panel, expand **11. Terrestrial Flora and Fauna.kmz** and then open the **11.4 Biomes** folder.

Instructions for Exploration 11.4 Biomes Parts A–B:

1. Open the **Biome Classification** folder.

2. Examine the landscapes at each of the five placemarks labeled *A* through *E*. Consider the location of each placemark and note any evidence of natural vegetation on the landscape.

Exploration 11.4 Biomes Part A

A. Which of the placemarks in the **Biome Classification** folder would be considered a mid-latitude grassland biome?

1. A
2. B
3. C
4. D
5. E

Exploration 11.4 Biomes Part B

B. Which of the placemarks in the **Biome Classification** folder would be considered a boreal forest biome?

1. A
2. B
3. C
4. D
5. E

When you complete the previous two parts, turn off and collapse the **Biome Classification** folder.

Instructions for Exploration 11.4 Biomes Part C:

1. Turn on and double-click the **Forest Change** folder. This folder displays data by country for the percent change in total forest extent from 2000 to 2005. Countries with a positive change in forest extent are shown in green, while countries with a negative change in forest extent are shown in red; countries with no change are in grey. The relative extrusion of each country from the surface of the globe indicates the amount of change, positive or negative.

2. Examine the patterns of forest change worldwide, noting the countries with the greatest increases and decreases in forest cover.

Exploration 11.4 Biomes Part C

C. Based on the data presented in the **Forest Change** folder, which of the following countries had the highest percentage increase in forest extent from 2000 to 2005?

1. Comoros
2. China
3. United States
4. Brazil
5. Iceland

When you complete this part, turn off the **Forest Change** folder.

Instructions for Exploration 11.4 Biomes Part D:

1. Turn on and double-click the **Rondônia, Brazil** folder. This folder contains imagery of part of the Brazilian province of Rondônia for 1975 and 2001.

2. Examine the details of the two image overlays and then compare the changes between 1975 and 2001 to the default Google Earth™ imagery.

3. Click the *Rondônia* hyperlink to learn more about the changes at this location.

Exploration 11.4 Biomes Part D

D. Which of the following statements is best supported by the imagery in the **Rondônia, Brazil** folder?

1. More than 50 percent of the world's tropical forests are in Brazil.
2. Since 2001, forest clearance has decreased sharply compared to earlier rates.
3. Fishbone development in the region occurs along new roads.
4. Forest clearing at this location is being done by a large multinational corporation.
5. The imagery suggests that the forest in the southeast part of the region is protected from removal by a natural barrier.

When you complete this exploration, turn off and collapse the **11.4 Biomes** folder.

Exploration 11.4 – Biomes
Short Answer

Instructions for all Parts:

1. Make sure you have opened the **KMZ** file from www.mygeoscienceplace.com.

2. From the Places panel, expand **11. Terrestrial Flora and Fauna.kmz** and then open the **11.4 Biomes** folder.

Instructions for Exploration 11.4 Biomes Short Answer A:

1. Double-click the **Santa Cruz, Bolivia** folder to fly to this region of eastern Bolivia. Examine the patterns of development that exist at this location.

2. Turn on and double-click the **Rondônia, Brazil** folder to fly to an area 850 km north-northwest and evaluate the patterns of development at that location.

Exploration 11.4 Biomes Short Answer A

A. Describe the differences between the patterns that exist in the **Santa Cruz, Bolivia** folder and those in the **Rondônia, Brazil** folder. You may utilize contemporary Google Earth™ imagery as well as the historical imagery available in each folder. Provide a hypothesis for why the landscapes of these locations appear different.

_____

_____

_____

_____

_____

_____

_____

When you complete this part, turn off and collapse the **Santa Cruz, Bolivia** and **Rondônia, Brazil** folders.

Instructions for Exploration 11.4 Biomes Short Answer B:

1. Use Google Earth™ to zoom to your location and find an area that is representative of the biome at your location.

Exploration 11.4 Biomes Short Answer B

B. Provide the latitude and longitude of a location near you that is representative of the biome you live in. What is this biome and what are the flora, fauna, and climate characteristics of this location that classify it as such?

_____

_____

_____

_____

_____

_____

_____

When you complete this exploration, turn off and collapse the **11.4 Biomes** folder.

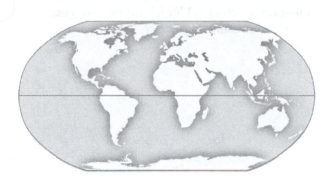

Encounter Physical Geography

Name:_____

Date: _____

Exploration 12.1 – Soil Forming Factors
Multiple Choice

Soil is a dynamic medium that is constantly evolving due to local geology, climate, topography, and biology over time. Through geographic analysis, we can identify which of these soil forming factors are dominant at one location compared to another.

Instructions for all Parts:

1. Make sure you have opened the **KMZ** file from www.mygeoscienceplace.com.

2. From the Places panel, expand **12. Soils**.kmz and then open the **12.1 Soil Forming Factors** folder.

3. Examine the five locations placemarked as *A* through *E* and evaluate each based on the soil forming factors as discussed in your textbook.

Exploration 12.1 Soil Forming Factors Part A

A. At which of the placemarks contained in the **Soil Forming Factors** folder would the biological factor of soil formation be limited due to a lack of organic matter?

1. A
2. B
3. C
4. D
5. E

Exploration 12.1 Soil Forming Factors Part B

B. At which of the placemarks contained in the **Soil Forming Factors** folder would the biological factor of soil formation be accelerated due to an abundance of organic matter?

1. A
2. B
3. C
4. D
5. E

Exploration 12.1 Soil Forming Factors Part C

C. At which of the placemarks contained in the **Soil Forming Factors** folder would the climactic factor retard soil development?

1. A
2. B
3. C
4. D
5. E

Exploration 12.1 Soil Forming Factors Part D

D. At which of the placemarks contained in the **Soil Forming Factors** folder would the topographic factor retard soil development?

1. A
2. B
3. C
4. D
5. E

When you complete this exploration, turn off and collapse the **12.1 Soil Forming Factors** folder.

Exploration 12.1 – Soil Forming Factors
Short Answer

Instructions for Exploration 12.1 Soil Forming Factors Short Answer A:

1. Make sure you have opened the **KMZ** file from www.mygeoscienceplace.com.

2. From the Places panel, expand **12. Soils.kmz** and then open the **12.1 Soil Forming Factors** folder.

3. Select one of the placemarked locations and double-click to zoom to the site. Evaluate the site, noting the presence or absence of soil forming factors.

Exploration 12.1 Soil Forming Factors Short Answer A

A. Hypothesize how at least three different soil forming factors have affected the site you selected from the **Soil Forming Factors** folder.

_____

_____

_____

_____

_____

_____

_____

Instructions for Exploration 12.1 Soil Forming Factors Short Answer B:

1. Use Google Earth™ to zoom to a location where you have witnessed widespread soil erosion. If you are unable to think of a specific site you have experienced, search "soil erosion location examples" on the Internet. Identify a specific location of erosion online, then zoom to that location in Google Earth™.

Exploration 12.1 Soil Forming Factors Short Answer B

B. Identify a location with widespread soil erosion by latitude and longitude. Hypothesize as to the local factors that might contribute to soil erosion at that location.

_____

_____

_____

_____

_____

_____

_____

When you complete this exploration, turn off and collapse the **12.1 Soil Forming Factors** folder.

12.2 – Soil Components and Composition
Multiple Choice

Soil varies based on the proportion of its primary components: inorganic material, organic material, air, and water. Those properties control processes such as infiltration and the ability of a soil to hold and maintain moisture.

Instructions for all Parts:

1. Make sure you have opened the **KMZ** file from www.mygeoscienceplace.com.

2. From the Places panel, expand **12. Soils.kmz** and then open the **12.2 Soil Components and Composition** folder.

Instructions for Exploration 12.2 Soil Components and Composition Part A:

1. Turn on and double-click the **Rio de Janeiro** folder.

2. Examine each of the five placemarked locations labeled *A* through *E*, noting any factors related to soil present at each site.

Exploration 12.2 Soil Components and Composition Part A

A. At which of the placemarks contained in the **Rio de Janeiro** folder is the parent material visible?

1. A
2. B
3. C
4. D
5. E

Exploration 12.2 Soil Components and Composition Part B

B. At which of the placemarks contained in the **Rio de Janeiro** folder would the ability of the soil to absorb water (percolation) be the highest?

1. A
2. B
3. C
4. D
5. E

When you complete the previous two parts, turn off and collapse the **Rio de Janeiro** folder.

Instructions for Exploration 12.2 Soil Components and Composition Part C:

1. Turn on and double-click the **Litter** folder.

2. Examine the five placemarks labeled *A* through *E*, noting the potential for plant-generated litter at each location.

Exploration 12.2 Soil Components and Composition Part C

C. At which of the placemarked locations in the **Litter** folder would the amount of plant-generated litter be highest?

1. A
2. B
3. C
4. D
5. E

When you complete this part, turn off and collapse the **Litter** folder.

Instructions for Exploration 12.2 Soil Components and Composition Part D:

1. Turn on and double-click the **Gravitational Water** folder. The data in this folder shows a radar image from thunderstorm activity in Louisiana on August 24, 2012.

2. Examine the radar data seen in the image and evaluate the patterns based on the potential for gravitational water at the time the data were captured. You may need to turn on the *Borders and Labels* layer from the *Primary Database* and use the Zoom and Navigation tools to provide more regional context for this thunderstorm event.

Exploration 12.2 Soil Components and Composition Part D

D. Based on the radar image associated with the **Gravitational Water** folder, at which of the following locations would you expect to find relatively higher rates of gravitational water?

1. Alexandria
2. Crowley
3. Beaumont
4. Sulfur
5. Silsbee

When you complete this exploration, turn off and collapse the **12.2 Soil Components and Composition** folder.

Exploration 12.2 – Soil Components and Composition
Short Answer

Instructions for all Short Answers

1. Make sure you have opened the **KMZ** file from www.mygeoscienceplace.com.

2. From the Places panel, expand **12. Soils.kmz** and then open the **12.2 Soil Components and Composition** folder.

Instructions for Exploration 12.2 Soil Components and Composition Short Answer A:

1. Double-click the **Rio de Janiero** folder.

2. Examine the urban scene presented here, being sure to note any factors related to soil and soil development.

Exploration 12.2 Soil Components and Composition Short Answer A

A. In general, what impacts would the urban features seen in the **Rio de Janeiro** folder have on the development of soil? Be sure to use specific examples from the imagery to explain the process of soil formation in an urban location.

_____

_____

_____

_____

_____

_____

_____

Instructions for Exploration 12.2 Soil Components and Composition Short Answer B:

1. Using Google Earth™, zoom to a location near you that would likely have the highest concentration of litter.

Exploration 12.2 Soil Components and Composition Short Answer B

B. Identify by latitude and longitude a location in your area that would likely have a high amount of litter. Explain why this location is a likely location for litter.

_____

_____

_____

_____

_____

_____

_____

When you complete this exploration, turn off and collapse the **12.2 Soil Components and Composition** folder.

12.3 – Soil Properties and Chemistry
Multiple Choice

Characteristics such as color, texture, structure, and chemical makeup enable humans to differentiate and classify different soils. This understanding empowers humans to maximize the use of this resource.

Instructions for all Parts:

1. Make sure you have opened the **KMZ** file from www.mygeoscienceplace.com.

2. From the Places panel, expand **12. Soils.kmz**.

3. Turn on the **SoilWeb** interactive folder. This folder accesses data from the US Department of Agriculture's National Cooperative Soil Survey network, hosted by the California Soil Resource Lab. Because this data comes from outside of Google Earth™, it may take a few moments to load.

4. Open the **12.3 Soil Properties and Chemistry** folder.

Instructions for Exploration 12.3 Soil Properties and Chemistry Part A:

1. Turn on and double-click the *Soil View 1* placemark. You should be zoomed to a location in Sherman County, Kansas. At this scale, the *SoilWeb* provides you with a map of local soil series classifications, indicated with yellow lines and labels.

2. Click on the *SoilWeb* label closest to the *Soil View 1* placemark (it should be labeled '1870'). Examine the resulting pop-up window for a more detailed description of the soil along with a generalized profile of soil in that location.

Exploration 12.3 Soil Properties and Chemistry Part A

A. Based on the information available through the **SoilWeb** folder data link adjacent to the *Soil View 1* placemark, which of the following statements is most strongly supported?

1. The Pleasant soils represent the greatest percentage of the Ulysses-Keith silt loams.
2. The Ulysses-Keith silt loams are found in locations with slopes of greater than 3 percent.
3. The A horizon of the Pleasant soil is typically deeper than the A horizon of the Ulysses or Keith soils.
4. The B horizon of the Pleasant soil is typically thicker than the B horizons of the Ulysses or Keith soils.
5. The Ulysses soils are usually found in the plains areas as opposed to the ridges or playas.

Instructions for Exploration 12.3 Soil Properties and Chemistry Part B:

1. Ensure that the pop-up information window for the *SoilWeb* label nearest the *Soil View 1* placemark is open. (From the "1870" label.)

2. Click the hyperlinked block diagram named *KS-2012-01-26-01* from the appropriate section in the pop-up. This will open a generalized sketch of the soils present in this part of Sherman County.

3. Examine the structure and distribution of soils at this location as shown in block diagram *KS-2012-01-26-01* from the *SoilWeb* data.

Exploration 12.3 Soil Properties and Chemistry Part B

B. Based on the information available through the block diagram associated with the **SoilWeb** folder data adjacent to the *Soil View 1* placemark, which of the following statements is most strongly supported?

1. In general, Ulysses soils are found at topographically higher locations than Keith soils.
2. Pleasant soils are found in areas of stream erosion.
3. Ulysses and Keith soils are not found adjacent to one another.
4. This diagram suggests that Keith soils are more extensive in this immediate region.
5. Alluvium is found along ridgelines.

When you complete the previous parts, close all pop-up windows and associated data and turn off the *Soil View 1* placemark.

Instructions for Exploration 12.3 Soil Properties and Chemistry Parts C–D:

1. Turn on and double-click the *Soil View 2* placemark. You are now zoomed to a location in Jackson County, Indiana. The *SoilWeb* data will provide you with a map of local soil series classifications, shown in yellow lines and labels.

2. Click on the *SoilWeb* label closest to the *Soil View 2* placemark (labeled "BbhA"). From the resulting pop-up window, click the Bartle hyperlink to open a page with additional soil characteristics of this site.

3. Examine the data about this soil, being sure to note the information provided in the diagrams at the bottom of the page.

Exploration 12.3 Soil Properties and Chemistry Part C

C. Based on the data found in the additional soil characteristics page for the *SoilWeb* link closest to the *Soil View 2* placemark, at which of the following depths is this soil the most acidic?

1. 0 cm
2. 51 cm
3. 102 cm
4. 152 cm
5. 203 cm

Exploration 12.3 Soil Properties and Chemistry Part D

D. Based on the data found in the additional soil characteristics page for the *SoilWeb* link closest to the *Soil View 2* placemark, which of the following statements is most strongly supported?

1. Based on the percentage of clay and sand in this soil, it is likely classified as a silty loam.
2. This soil has a high concentration of calcium carbonate.
3. This soil is well drained.
4. The B horizon begins at 10 cm below the surface of a typical profile.
5. Eastern poison ivy does not grow on this soil due to high gypsum content.

When you complete this exploration, close any pop-up or additional windows, turn off the **SoilWeb** folder, then turn off and collapse the **12.3 Soil Properties and Chemistry** folder.

Exploration 12.3 – Soil Properties and Chemistry
Short Answer

Instructions for all Parts:

1. Make sure you have opened the **KMZ** file from www.mygeoscienceplace.com.

2. From the Places panel, expand **12. Soils.kmz**.

3. Turn on the **SoilWeb** interactive folder. This folder accesses data from the US Department of Agriculture's National Cooperative Soil Survey network, hosted by the California Soil Resource Lab. Because this data comes from outside of Google Earth™, it may take a few moments to load.

4. Open the **12.3 Soil Properties and Chemistry** folder.

Instructions for Exploration 12.3 Soil Properties and Chemistry Short Answer A:

1. Turn on and double-click the *Soil View 1* placemark.

2. Examine the soil complexity of the surrounding region as displayed by the *SoilWeb* data showing soil series classifications.

3. Repeat the same examination at the location of the *Soil View 2* placemark.

Exploration 12.3 Soil Properties and Chemistry Short Answer A

A. Compare the soil complexity patterns found in the soil series at the *Soil View 1* and *Soil View 2* placemarks. Provide an explanation for the relative levels of complexity found at each location.

_____
_____
_____
_____
_____
_____
_____

Instructions for Exploration 12.3 Soil Properties and Chemistry Short Answer B:

1. Zoom to your location in Google Earth™, making sure to view the area at a scale where the soil series classification (yellow lines and labels) is visible.

2. Locate a nearby non-urban complex soil series and examine the detailed information about that soil by clicking on the appropriate links. (If the soil found in your immediate area is classified as urban, find the nearest non-urban soil classification for evaluation.)

Exploration 12.3 Soil Properties and Chemistry Short Answer B

B. Name and briefly describe the soil for your location. Be sure to indicate any significant characteristics, soil properties, and profile structure information, as well as the latitude and longitude of your location.

_____
_____
_____
_____
_____
_____
_____

When you complete this exploration, turn off the **SoilWeb** folder then turn off and collapse the **12.3 Soil Properties and Chemistry** folder.

Exploration 12.4 – Soil Classification
Multiple Choice

Soil classification can be a complicated endeavor and several different methods exist to define regional soil patterns. A simple type of classification focuses on the primary soil creation, or pedogenic, regime while other classifications encompass a broad array of criteria including climate, mineral composition, and acidity.

Instructions for all Parts:

1. Make sure you have opened the **KMZ** file from www.mygeoscienceplace.com.

2. From the Places panel, expand **12. Soils.kmz** then open the **12.4 Soil Classification** folder.

3. Examine each of the five placemarks, paying particular attention to environmental contexts such as the climate or vegetative regimes present at each location.

Exploration 12.4 Soil Classification Part A

A. At which of the placemarked locations in the **12.4 Soil Classification** folder would salinization be the dominant pedogenic regime?

1. A
2. B
3. C
4. D
5. E

Exploration 12.4 Soil Classification Part B

B. At which of the placemarked locations in the **12.4 Soil Classification** folder would gleization be the dominant pedogenic regime?

1. A
2. B
3. C
4. D
5. E

Exploration 12.4 Soil Classification Part C

C. Based on the landscape characteristics near each placemark in the **12.4 Soil Classification** folder and information from your textbook, which of the placemarks is located in an area with Ultisol soils?

1. A
2. B
3. C
4. D
5. E

Exploration 12.4 Soil Classification Part D

D. Based on the landscape characteristics near each placemark in the **12.4 Soil Classification** folder and information from your textbook, which of the placemarks is located in an area with Spodosol soils?

1. A
2. B
3. C
4. D
5. E

When you complete this exploration, turn off and collapse the **12.4 Soil Classification** folder.

Exploration 12.4 – Soil Classification
Multiple Choice

Instructions for all Parts:

1. Make sure you have opened the **KMZ** file from www.mygeoscienceplace.com.

2. From the Places panel, expand **12. Soils.kmz**.

Instructions for Exploration 12.4 Soil Classification Short Answer A:

1. Turn on the **SoilWeb** interactive folder. This folder accesses data from the US Department of Agriculture's National Cooperative Soil Survey network, hosted by the California Soil Resource Lab. Because this data comes from outside of Google Earth™, it may take a few moments to load.

Exploration 12.4 Soil Classification Short Answer A

A. Using the data available in the *SoilWeb* folder, describe the soil taxonomy at your location. Include the order, suborder, greatgroup, subgroup, family, and soil series. Describe the pedogenic regime that is most responsible for providing the soil geography of your area. Be sure to note your location in your response.

_____
_____
_____
_____
_____
_____
_____

Instructions for Exploration 12.4 Soil Classification Short Answer B:

1. Open the **12.4 Soil Classification** folder.

2. Examine each of the five placemarked locations in conjunction with your text to identify the dominant soil order represented by each placemark.

Exploration 12.4 Soil Classification Short Answer B

B. Identity and list the five soil orders represented by each of the placemarks in the **12.4 Soil Classification** folder. Use outside resources to identify the soil order that underlays much of the productive grain-producing areas of the United States. Provide an appropriate latitude and longitude for a location where this soil order is found.

_____
_____
_____
_____
_____
_____
_____

When you complete this exploration, turn off and collapse the **12.4 Soil Classification** folder.

Encounter Physical Geography

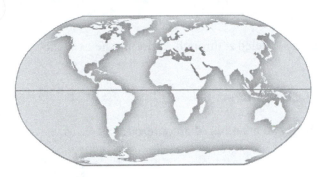

Name:_____

Date: _____

Exploration 13.1 – Structure and Composition of the Earth I
Multiple Choice

Geographers are interested in the basic building blocks that constitute landforms found across the Earth's surface. The unique topographic configurations are a result of the Earth's minerals and rocks being assembled and disassembled by the internal and external processes of our planet.

Instructions for all Parts:

1. Make sure you have opened the **KMZ** file from www.mygeoscienceplace.com.

2. From the Places panel, expand **13. Introduction to Landform Study.kmz** and then open the **13.1 Structure and Composition of the Earth I** folder.

Instructions for Exploration 13.1 – Structure and Composition of the Earth I Part A:

1. Open and turn on the **Crust** folder.

2. Examine each of the five placemarks labeled *A* through *E*, noting the likely characteristics of the Earth's crust at each location.

Exploration 13.1 Structure and Composition of the Earth I Part A

A. At which of the placemarks in the **Crust** folder would you expect the crust of the Earth to be thinnest?

1. A
2. B
3. C
4. D
5. E

When you complete this question, turn off and collapse the **Crust** folder.

Instructions for Exploration 13.1 Structure and Composition of the Earth I Part B:

1. Open and turn on the **Outcrop** folder.

2. Examine each of the five placemarked locations labeled *A* through *E*, noting the existence of exposed rock at each location.

Exploration 13.1 Structure and Composition of the Earth I Part B

B. Which of the placemarks in the **Outcrop** folder is **not** an example of an outcrop?

1. A
2. B
3. C
4. D
5. E

When you complete this question, turn off and collapse the **Outcrop** folder.

Instructions for Exploration 13.1 Structure and Composition of the Earth I Part C:

1. Turn on and double-click the *Kizimen* placemark. This location marks a prominent volcanic mountain on Russia's Kamchatka Peninsula.

Exploration 13.1 Structure and Composition of the Earth I Part C

C. Based on the evidence found at the *Kizimen* placemark, which of the following would you be **least** likely to encounter at this location?

1. magma
2. igneous rocks
3. lava
4. pyroclastics
5. basalt

Instructions for Exploration 13.1 Structure and Composition of the Earth I Part D:

1. Turn on and double-click the *Kaibab Trail* Gigapan image. This panoramic photo provides a detailed glimpse into a part of the Grand Canyon.

2. Examine the photo, paying close attention to the structure of rocks at this location.

Exploration 13.1 Structure and Composition of the Earth I Part D

D. Based on the evidence in the *Kaibab Trail* Gigapan image, what type of rock is found in this part of the Grand Canyon?

1. clastic sedimentary
2. extrusive igneous
3. metamorphic
4. organic sedimentary
5. intrusive igneous

When you complete this exploration, turn off and collapse the **13.1 Structure and Composition of the Earth I** folder.

Exploration 13.1 Structure and Composition of the Earth I
Short Answer

Instructions for all Parts:

1. Make sure you have opened the **KMZ** file from www.mygeoscienceplace.com.

2. From the Places panel, expand **13. Introduction to Landform Study.kmz** and then open the **13.1 Structure and Composition of the Earth I** folder.

Instructions for Exploration 13.1 Structure and Composition of the Earth I
Short Answer A:

1. Open the **Crust** folder. Examine each of the five placemarks labeled *A* through *E*, being sure to consider the status of the Earth's crust at each location.

Exploration 13.1 Structure and Composition of the Earth I Short Answer A

A. If you were attempting to drill a hole through the crust, at which of the five placemarks in the **Crust** folder would conditions be the most ideal? Be sure to consider both the relative thickness of the crust as well as other factors relevant to deep-drilling operations.

_____

_____

_____

_____

_____

_____

_____

When you complete this question, turn off and collapse the **Crust** folder.

**Instructions for Exploration 13.1 Structure and Composition of the Earth I
Short Answer B:**

1. Use Google Earth™ to navigate to your location.

2. Examine the landscape of your immediate area and identify the closest rock outcropping.

Exploration 13.1 Structure and Composition of the Earth I Short Answer B

B. Identify a rock outcropping near your location by latitude and longitude. Provide landmarks for context and explain any local significance this site may have.

_____
_____
_____
_____
_____
_____
_____

When you complete this exploration, turn off and collapse the **13.1 Structure and Composition of the Earth I** folder.

**Exploration 13.2 – Structure and Composition of the Earth II
Multiple Choice**

Rocks can be formed below, within, and above the Earth's crust. As forces such as pressure and heat act alongside weathering and erosional processes at the surface, three major rock classes are identified: igneous, sedimentary, and metamorphic.

Instructions for all Parts:

1. Make sure you have opened the **KMZ** file from www.mygeoscienceplace.com.

2. From the Places panel, expand **13. Introduction to Landform Study.kmz** and then open the **13.2 Structure and Composition of the Earth II** folder.

Instructions for Exploration 13.2 Structure and Composition of the Earth II Part A:

1. Turn on and double-click the _Mt. Etna_ placemark.

2. Examine the landscape at this location and evaluate the likely type of rocks present.

Exploration 13.2 Structure and Composition of the Earth II Part A

A. Based on the available evidence at the *Mt. Etna* placemark, what type of rock would you likely find at this location?

1. Extrusive igneous
2. Metamorphic
3. Clastic sedimentary
4. Organic sedimentary
5. Intrusive igneous

Instructions for Exploration 13.2 Structure and Composition of the Earth II Part B:

1. Turn on and double-click the *Cancun* placemark.

2. Examine the landscape at this location, being sure to note the presence of any rock-forming minerals.

Exploration 13.2 Structure and Composition of the Earth II Part B

B. Based on the available evidence at the *Cancun* placemark, what mineral group is most common at this location?

1. silicates
2. sulfides
3. oxides
4. halides
5. carbonates

Instructions for Exploration 13.2 Structure and Composition of the Earth II Part C:

1. Turn on and double-click the *Shatt al-Arab* placemark.

2. Examine the landscape at this location, paying particular attention to the rock-forming processes present.

Exploration 13.2 Structure and Composition of the Earth II Part C

C. Based on the available evidence at the *Shatt al-Arab* placemark, what type of rock is most likely being formed here?

1. Clastic sedimentary
2. Extrusive igneous
3. Metamorphic
4. Organic sedimentary
5. Intrusive igneous

Instructions for Exploration 13.2 Structure and Composition of the Earth II Part D:

1. Turn on and double-click the *Congo Basin* placemark.

2. Examine the landscape at this location, noting the rock-forming processes likely present here.

Exploration 13.2 Structure and Composition of the Earth II Part D

D. Based on the available evidence at the *Congo Basin* placemark, what type of rock is most likely being formed here?

1. Organic sedimentary
2. Clastic sedimentary
3. Extrusive igneous
4. Metamorphic
5. Intrusive igneous

When you complete this exploration, turn off and collapse the **13.2 Structure and Composition of the Earth II** folder.

Exploration 13.2 – Structure and Composition of the Earth II
Short Answer

Instructions for Exploration 13.2 Structure and Composition of the Earth II
Short Answer A:

1. Navigate to mrdata.usgs.gov/geology/state using your web browser to access the *Geologic Maps of US States – Spatial Data* website. Click on your state from the available list.

2. From the *View* sub-heading, click on the KMZ link to download the Google Earth™ file for your state. Open this file in Google Earth™ and then zoom to your location.

3. Click on the map at your location to open a pop-up window with a link to a more detailed description of the local geology. Click the link for a detailed description of the soils at your location.

Exploration 13.2 – Structure and Composition of the Earth II Short Answer A

A. Using the information gathered from the Google Earth™ file downloaded from the USGS, describe the local geology at your location. Be sure to note your location as well as the rock types present, the name of the major rock layer, the rock age, and any significant characteristics.

_____
_____
_____
_____
_____
_____
_____

When you complete this question, close any pop-up windows or external websites associated with layers added from the USGS.

Instructions for Exploration 13.2 Structure and Composition of the Earth II Short Answer B:

1. Make sure you have opened the **KMZ** file from www.mygeoscienceplace.com.

2. From the Places panel, expand **13. Introduction to Landform Study.kmz** and then open the **13.2 Structure and Composition of the Earth II** folder. Examine the four placemarked sites, paying particular attention to any potential examples from the rock cycle present at each location.

Exploration 13.2 Structure and Composition of the Earth II Short Answer B

B. Using examples from the locations contained in the **13.2 Structure and Composition of the Earth II** folder, describe significant portions of the rock cycle.

_____

_____

_____

_____

_____

_____

_____

When you complete this exploration, turn off and collapse the **13.2 Structure and Composition of the Earth II** folder.

Exploration 13.3 – Study of Landforms
Multiple Choice

Internal and external forces of the Earth act upon different rock types to create the range of landforms that appear on the surface. These landforms create high and low points of relief, on a range of scales from local to regional to global.

Instructions for all Parts:

1. Make sure you have opened the **KMZ** file from www.mygeoscienceplace.com.

2. From the Places panel, expand **13. Introduction to Landform Study.kmz** and then open the **13.3 Study of Landforms** folder.

Instructions for Exploration 13.3 Study of Landforms Part A:

1. Open the **Scale** folder.

2. Double-click on each of the five placemarked locations labeled *1* through *5,* noting the relative scale of each placemark.

A. Exploration 13.3 Study of Landforms Part A

A. Based on the relative position of the placemarks in the **Scale** folder, what is the correct order of view from largest to smallest scale?

1. 3, 4, 1, 2, 5
2. 1, 2, 3, 4, 5
3. 5, 4, 3, 2, 1
4. 5, 2, 1, 4, 3
5. 2, 5, 3, 1, 4

When you complete this question, turn off and collapse the **Scale** folder.

Instructions for Exploration 13.3 Study of Landforms Part B:

1. Open the **Relief** folder.

2. Double-click each of the *Scene* placemarks labeled *A* through *E*, being sure to note the amount of total relief present in each location.

Exploration 13.3 Study of Landforms Part B

B. Based on the evidence at the placemarks in the **Relief** folder, which of the placemarked locations has the greatest amount of relief?

1. A
2. B
3. C
4. D
5. E

Instructions for Exploration 13.3 Study of Landforms Parts C–D:

1. Open the **Internal and External Processes** folder.

2. Double-click each of the placemarks, noting evidence of internal or external processes in the surface landscapes at each location.

Exploration 13.3 Study of Landforms Part C

C. Based on landscape characteristics, which of the locations in the **Internal and External Processes** folder are more directly a result of internal processes?

1. Glacial, Waves, Folding
2. Intrusive volcanism, Folding
3. Waves, Folding
4. Fluvial, Glacial, Waves
5. Intrusive volcanism, Waves, Folding

Exploration 13.3 Study of Landforms Part D

D. Based on landscape characteristics, which of the locations in the **Internal and External Processes** folder are more directly a result of external processes?

1. Glacial, Waves, Folding
2. Intrusive volcanism, Folding
3. Waves, Folding
4. Fluvial, Glacial, Waves
5. Intrusive volcanism, Waves, Folding

When you complete this exploration, turn off and collapse the **13.3 Study of Landforms** folder.

Exploration 13.3 – Study of Landforms
Short Answer

Instructions for Exploration 13.3 Study of Landforms Short Answer A:

1. Make sure you have opened the **KMZ** file from www.mygeoscienceplace.com.

2. From the Places panel, expand **13. Introduction to Landform Study.kmz** and then open the **13.3 Study of Landforms** folder.

3. Open the **Scale** folder. Compare the different views represented by the placemarks labeled *A* through *E*, being sure to note the relative scale at each location.

Exploration 13.3 Study of Landforms Short Answer A

A. Indicate the placemarks that represent the largest and smallest scales. List and discuss at least one benefit and one drawback of using large-scale and small-scale maps in analysis.

_____
_____
_____
_____
_____
_____
_____

Instructions for Exploration 13.3 Study of Landforms Short Answer B:

1. Use Google Earth™ to zoom to the level of your county or equivalent sub-state administrative jurisdiction.

2. Examine the physical landscapes of your county, noting significant features as well as the relative high and low points. Remember to use the Look Around navigation control to tilt your view and to check the elevation for the cursor location by checking the status bar at the bottom of the screen.

Exploration 13.3 Study of Landforms Short Answer B

B. What is the total amount of relief present in your county? Provide some possible internal or external processes that may have contributed to the landscapes in your county. Be sure to indicate the name of your county.

_____

_____

_____

_____

_____

_____

_____

When you complete this exploration, turn off and collapse the **13.3 Study of Landforms** folder.

Exploration 13.4 –Landform Assemblages
Multiple Choice

Physical landforms can be organized and classified into distinct assemblages based on topographical characteristics such as relief and dominant features. By defining and assigning landform assemblages, we can compare and contrast the physical landscapes of unique places across the globe.

Instructions for all Parts:

1. Make sure you have opened the **KMZ** file from www.mygeoscienceplace.com.

2. From the Places panel, expand **13. Introduction to Landform Study**.kmz and then open the **13.4 Landform Assemblages** folder.

3. Open the **Examples of Landform Assemblages** folder. Each of the placemarks in this folder identifies one of seven major landform assemblages around the world. Examine each of these sites noting the characteristics of the physical landscapes at each site.

4. Open the **Identify** folder. Examine each of the five placemarked locations labeled *A* through *E*, being sure to note any similarities to the sample locations.

Exploration 13.4 Landform Assemblages Part A

A. Based on the evidence at the sample placemarks in the **Examples of Landform Assemblages** folder, which of the placemarks in the **Identity** folder best exemplifies the Hills landform assemblage?

    1. A
    2. B
    3. C
    4. D
    5. E

Exploration 13.4 Landform Assemblages Part B

B. Based on the evidence at the sample placemarks in the **Examples of Landform Assemblages** folder, which of the placemarks in the **Identity** folder best exemplifies the tablelands landform assemblage?

    1. A
    2. B
    3. C
    4. D
    5. E

Exploration 13.4 Landform Assemblages Part C

C. Based on the evidence at the sample placemarks in the **Examples of Landform Assemblages** folder, which of the placemarks in the **Identity** folder best exemplifies the flat plains landform assemblage?

    1. A
    2. B
    3. C
    4. D
    5. E

Exploration 13.4 Landform Assemblages Part D

D. Based on the evidence at the sample placemarks in the **Examples of Landform Assemblages** folder, which of the placemarks in the **Identity** folder best exemplifies the Mountains landform assemblage?

    1. A
    2. B
    3. C
    4. D
    5. E

When you complete this exploration, turn off and collapse the **13.4 Landform Assemblages** folder.

Exploration 13.4 – Landform Assemblages
Short Answer

Instructions for Exploration 13.4 Landform Assemblages Short Answer A:

1. Use Google Earth™ to zoom to the level of your county or equivalent sub-state administrative jurisdiction.

2. Examine the physical landscapes of your county, noting significant patterns that reveal the dominant landform assemblage of the area.

3. Using information from your text as well as landscape evidence in Google Earth™, find a location on another continent that shares the same landform assemblage as your county.

Exploration 13.4 Landform Assemblages Short Answer A

A. What landform assemblage is the most dominant in your county? Be certain to indicate your location. Identify by name and latitude and longitude a location on a different continent that has a similar landscape, based on landform assemblages.

_____

_____

_____

_____

_____

_____

_____

Instructions for Exploration 13.4 Landform Assemblages Short Answer B:

1. Make sure you have opened the **KMZ** file from www.mygeoscienceplace.com.

2. From the Places panel, expand **13. Introduction to Landform Study.kmz** and then open the **13.4 Landform Assemblages** folder.

3. Open the **Examples of Landform Assemblages** folder. Select one of the sample locations and evaluate the landscape at that site, paying particular attention to the processes that helped form that landscape.

Exploration 13.4 Landform Assemblages Short Answer B

B. Name the landform assemblage you have selected. Describe the internal and external processes that contributed to the creation and subsequent modification of the landscape at this location.

_____

_____

_____

_____

_____

_____

_____

When you complete this exploration, turn off and collapse the **13.4 Landform Assemblages** folder.

Encounter Physical Geography

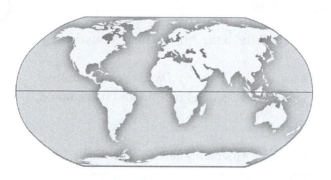

Name:_____

Date: _____

Chapter 14
Internal Processes

Exploration 14.1 – Plate Tectonics I
Multiple Choice

While the positions of the world's continents may seem static, the continental landmasses are actually in a state of perpetual movement. Over millennia, these landmasses have been torn apart, collided with one another, and evolved in a variety of configurations. An understanding of the underlying principles of this movement provides a contextual framework for studying tectonic landforms and hazards.

Instructions for all Parts:

1. Make sure you have opened the **KMZ** file from www.mygeoscienceplace.com.

2. From the Places panel, expand **14. Internal Processes.kmz** and then open the **14.1 Plate Tectonics I** folder.

Instructions for Exploration 14.1 Plate Tectonics I Part A:

1. Turn on and open the **Pangaea** folder. The placemarks in this folder are arranged as pairs of locations labeled *A* through *E*.

2. Examine the two locations for each of the five pairs, being sure to note the relationship between paired points in current geologic time as well as their expected positions during the time of the super-continent Pangaea.

Exploration 14.1 Plate Tectonics I Part A

A. Which of the labeled placemark pairs in the **Pangaea** folder represents locations that would have been adjacent or nearly adjacent when the super-continent was contiguous?

1. A
2. B
3. C
4. D
5. E

When you complete this part, turn off and collapse the **Pangaea** folder.

Instructions for Exploration 14.1 Plate Tectonics I Part B:

1. Use the search panel in Google Earth™ to locate and fly to the Appalachian Mountains.

2. Ensure that the *Borders and Labels* item is checked in the *Primary Database*, then mouse over the green label for the Appalachian Mountains. The general outline for this geologic region will appear.

3. Examine the generalized extent of the Appalachian Mountains paying attention to the potential geologic links of this region to other mountainous areas. If necessary consult your text or outside resources for information about the geologic history of the Appalachians.

Exploration 14.1 Plate Tectonics I Part B

B. Based on the evidence found in Google Earth™, which of the following mountain ranges are most closely linked geologically to the Appalachian Mountains?

1. Zagros Mountains
2. Cascade Range
3. Grampian Mountains
4. Rocky Mountains
5. Drakensburg Mountains

Instructions for Exploration 14.1 Plate Tectonics I Parts C–D:

1. Turn on and double-click the **Bathymetry** folder. The image overlay shows the depths of the ocean basins in meters below sea level where lighter colors indicate shallower water and darker colors indicate deeper water.

2. Examine each of the five placemarks labeled *1* through *5*, which are located at various points throughout the North Atlantic basin. Be sure to take note of the ocean terrain and relevant features at each location.

Exploration 14.1 Plate Tectonics I Part C

C. Based on the information in the **Bathymetry** folder, which of the following placemarks is located along a mid-ocean ridge?

1. A
2. B
3. C
4. D
5. E

Exploration 14.1 Plate Tectonics I Part D

D. Based on the information in the **Bathymetry** folder, at which placemark would you expect to find the oldest oceanic crust material?

1. A
2. B
3. C
4. D
5. E

When you complete this exploration, turn off and collapse the **14.1 Plate Tectonics I** folder.

Exploration 14.1 – Plate Tectonics I
Short Answer

Instructions for Exploration 14.1 Plate Tectonics I Short Answer A:

1. Use the search panel to locate and zoom to the Galapagos Islands. Examine the islands and the context of their surrounding region.

Exploration 14.1 Plate Tectonics I Short Answer A

A. Based on your knowledge of plate tectonics and seafloor spreading, in what direction would you expect the Galapagos Islands to be moving? Be certain to explain your response using the proper geologic terminology.

_____

_____

_____

_____

_____

_____

_____

Instructions for Exploration 14.1 Plate Tectonics I Short Answer B:

1. Make sure you have opened the **KMZ** file from www.mygeoscienceplace.com.

2. From the Places panel, expand **14. Internal Processes.kmz** and then open the **14.1 Plate Tectonics I** folder.

3. Turn on the **Bathymetry** folder.

4. Examine the image overlay showing the depth of the ocean basins in meters below sea level, being certain to note the scale used to display the data.

Exploration 14.1 Plate Tectonics I Short Answer B

B. Citing examples from the imagery in the **Bathymetry** folder, explain the lighter-colored areas that surround most continents. Be sure to note why these features are large off of some coasts and nearly non-existent off of others.

_____

_____

_____

_____

_____

_____

_____

When you complete this exploration, turn off and collapse the **14.1 Plate Tectonics I** folder.

Exploration 14.2 – Plate Tectonics II
Multiple Choice

The ongoing movement of the Earth's tectonic plates is responsible for prominent landforms and potentially disastrous natural hazards. Mountain ranges, volcanoes, and earthquakes are some of the phenomena related to tectonic action.

Instructions for all Parts:

1. Make sure you have opened the **KMZ** file from www.mygeoscienceplace.com.

2. From the Places panel, expand **14. Internal Processes.kmz** and then open the **14.2 Plate Tectonics II** folder.

Instructions for Exploration 14.2 Plate Tectonics II Parts A–C:

1. Turn on the _Earth's Tectonic Plates_ overlay. The lines on the map indicate major plate boundaries across the globe.

2. Examine the distribution of different plate boundary types, being sure to notice significant patterns relative to landmasses.

Exploration 14.2 Plate Tectonics II Part A

A. Based on the proximity to transform or convergent boundaries in the _Earth's Tectonic Plates_ overlay, what world city is **least** likely to be affected by an earthquake?

1. Tokyo
2. Berlin
3. Islamabad
4. Guatemala City
5. Istanbul

Exploration 14.2 Plate Tectonics II Part B

B. Based on the evidence in the *Earth's Tectonic Plates* overlay, which of the following mountain ranges is most likely to have volcanic activity due to its proximity to a convergent plate boundary?

1. Great Dividing Range
2. Ural Mountains
3. Tien Shan
4. Cascade Range
5. Appalachian Mountains

Exploration 14.2 Plate Tectonics II Part C

C. Based on the evidence in the *Earth's Tectonic Plates* overlay, which of the following statements is most strongly supported?

1. The largest plate is the Caribbean Plate.
2. Though rare, divergent boundaries can be found on land.
3. The Mediterranean region is not at risk for earthquakes or volcanic activity due to convergent boundaries.
4. The continent of Australia has the most complex collection of plate boundaries of all continental land masses.
5. Most convergent boundaries occur in the middle of continents.

When you complete Part C, turn off the *Earth's Tectonic Plates* overlay.

Instructions for Exploration 14.2 Plate Tectonics II Part D

1. Turn on the **Historic Large Earthquakes** folder. The points shown in this folder indicate the location of earthquakes of 7 or higher magnitude that have occurred since 1970. This data comes from the United States Geological Survey and may take a moment to load. Clicking on any of the data points will open a pop-up window containing the earthquake's magnitude, location, date, and depth.

2. Ensure that the *Earth's Tectonic Plates* overlay is turned on, then examine the distribution of large earthquakes in the context of the major plate boundaries.

Exploration 14.2 Plate Tectonics II Part D

D. Based on the data in the **Historic Large Earthquakes** folder, which of the following plate boundaries has been most active in terms of producing large earthquakes?

1. The boundary between the Nazca and Pacific plates.
2. The boundary between the Eurasian and North American plates.
3. The boundary between the Arabian and African plates.
4. The boundary between the Pacific and Australian plates.
5. The boundary between the Antarctic and Scotia plates.

When you complete this exploration, turn off and collapse the **14.2 Plate Tectonics II** folder.

Exploration 14.2 – Plate Tectonics II
Short Answer

Instructions for all Parts:

1. Make sure you have opened the **KMZ** file from www.mygeoscienceplace.com.

2. From the Places panel, expand **14. Internal Processes.kmz** and then open the **14.2 Plate Tectonics II** folder.

Instructions for Exploration 14.2 Plate Tectonics II Short Answer A:

1. Turn on the **Historic Large Earthquakes** folder. The points shown in this folder indicate the location of earthquakes of 7 or higher magnitude that have occurred since 1970. This data comes from the United States Geological Survey and may take a moment to load.

2. Evaluate the location of large earthquakes and note the regions where earthquake activity is either highly concentrated or extremely limited.

Exploration 14.2 Plate Tectonics II Short Answer A

A. If you were looking for a place to live where the risk of a very large earthquake was minimal, what location would you consider best? Be sure to use the appropriate terminology in your response.

_____
_____
_____
_____
_____
_____
_____

Instructions for Exploration 14.2 Plate Tectonics II Short Answer B:

1. Ensure that the **Historic Large Earthquake** folder is on, and then turn on the *Earth's Tectonic Plates* overlay.

2. Use the data in Google Earth™ and any other source to identify the locations of the largest earthquakes in the world, as presented by this dataset.

3. Examine the relationship between large earthquakes and the location of plate boundaries.

Exploration 14.2 Plate Tectonics II Short Answer B

B. Describe the correlation between large earthquake events and plate boundaries. Indicate any general patterns and explain the reasons for these patterns.

_____
_____
_____
_____
_____
_____
_____
_____

When you complete this exploration, turn off and collapse the **14.2 Plate Tectonics II** folder.

Exploration 14.3 – Volcanism
Multiple Choice

Volcanoes are perhaps the most dramatic surface representation of the Earth's tectonic forces. The uniquely identifiable landforms and intense active eruptions, alongside the ever-present risk to nearby human populations, make volcanoes a tremendously interesting focus of study.

Instructions for all Parts:

1. Make sure you have opened the **KMZ** file from www.mygeoscienceplace.com.

2. From the Places panel, expand **14. Internal Processes**.kmz and then open the **14.3 Volcanism** folder.

Instructions for Exploration 14.3 Volcanism Part A:

1. Turn on and open the **Eruption of Shiveluch Volcano** folder then turn on and double-click the *Overlay On* option. This image shows the eruption of a volcano on Russia's Kamchatka Peninsula.

2. Examine the image of this volcanic eruption, noting the characteristics of the material being released.

Exploration 14.3 Volcanism Part A

A. Based on the evidence in the **Eruption of Shiveluch Volcano** folder, which of the following terms associated with volcanism best describes the grey plume emanating from the volcano?

1. pyroclastic material
2. magma
3. lava flow
4. caldera
5. lahar

When you complete this part, turn off and collapse the **Eruption of Shiveluch Volcano** folder.

Instructions for Exploration 14.3 Volcanism Part B:

1. Turn on and open the **Shield or Composite** folder. Zoom to each of the five placemarked volcanoes.

2. Examine the volcanoes at each placemark, noting their shape and structure, being sure to consider the general characteristics of shield or composite volcanoes.

Exploration 14.3 Volcanism Part B

B. Based on the evidence seen in Google Earth™, which of the volcanoes placemarked in the **Shield or Composite** folder would be classified as shield volcanoes based on their form?

1. Mt. Fuji, Mt. Hood, and Mt. Taranaki
2. Mt. Fuji and Mauna Loa
3. Mt. Taranaki and Skjalberiour
4. Skjalberiour and Mauna Loa
5. Skjalberiour, Mt. Hood, and Mauna Loa

When you complete this part, turn off and collapse the **Shield or Composite** folder.

Instructions for Exploration 14.3 Volcanism Part C:

1. Double-click the *Shiprock* placemark to fly to this geologic feature in northwest New Mexico.

2. Examine the landscape found near Shiprock, noting the presence of volcanic landforms.

Exploration 14.3 Volcanism Part C

C. What two volcanic landforms are most apparent in the view provided at the *Shiprock* placemark?

1. pipe and laccolith
2. volcanic neck and dike
3. composite volcano and dike
4. shield volcano and dike
5. cinder cone and caldera

Instructions for Exploration 14.3 Volcanism Part D:

1. Double-click the *Crater Lake* placemark to fly to this geologic feature in southern Oregon.

2. Examine the landscapes found near Crater Lake, noting the presence of volcanic landforms.

Exploration 14.3 Volcanism Part D

D. What two volcanic landforms are most apparent in the view provided at the *Crater Lake* placemark?

1. pipe and laccolith
2. volcanic neck and dike
3. composite volcano and dike
4. shield volcano and dike
5. cinder cone and caldera

When you complete this exploration, turn off and collapse the **14.3 Volcanism** folder.

Exploration 14.3 – Volcanism
Short Answer

Instructions for all Parts:

1. Make sure you have opened the **KMZ** file from www.mygeoscienceplace.com.

2. From the Places panel, expand **14. Internal Processes.kmz** and then open the **14.3 Volcanism** folder.

Instructions for Exploration 14.3 Volcanism Short Answer A:

1. Turn on and double-click the *Mount St. Helens* Gigapan image.

2. Examine the landscape at this site, paying particular attention to any evidence of prior volcanic eruptions at this location.

Exploration 14.3 Volcanism Short Answer A

A. Describe at least two pieces of evidence seen in the *Mount St. Helens* Gigapan image that document the occurrence of a massive volcanic eruption in the recent past.

_____
_____
_____
_____
_____
_____
_____

Instructions for Exploration 14.3 Volcanism Short Answer B:

1. From the View menu, select Explore then Mars. This will switch the Google Earth™ 3D View to a perspective on Mars.

2. From the **14.3 Volcanism** folder, turn on and double-click the *Tharsis Tholus* placemark.

3. Explore this Martian volcano using the imagery in Google Earth™ and any other sources of information.

Exploration 14.3 Volcanism Short Answer B

B. Based on the imagery of Mars at the *Tharsis Tholus* placemark and information from outside sources, classify this and other Martian volcanoes as either shield or composite volcanoes. Be sure to explain why this classification is appropriate.

_____

_____

_____

_____

_____

_____

_____

IMPORTANT: After this question, be sure to use the View > Explore menu to change the 3D Viewer back to Earth.

When you complete this exploration, turn off and collapse the **14.3 Volcanism** folder.

Exploration 14.4 – Folding and Faulting
Multiple Choice

When the Earth's crust is subject to various types of stresses, it can result in deformation or breaking of crustal rocks. Referred to as folding and faulting in this context, these processes provide additional types of physical landforms that can be studied in Google Earth™.

Instructions for all Parts

1. Make sure you have opened the **KMZ** file from www.mygeoscienceplace.com.

2. From the Places panel, expand **14. Internal Processes.kmz** and then open the **14.4 Folding and Faulting** folder.

Instructions for Exploration 14.4 Folding and Faulting Part A:

1. Turn on and double-click the *Sideling Hill* placemark.

2. Examine the structure of the rock strata found in the road cut at this location.

Exploration 14.4 Folding and Faulting Part A

A. What type of geologic faulting and folding structure is revealed by the rock layers in the road cut at the *Sideling Hill* placemark?

1. syncline
2. anticline
3. overturned fold
4. horst and graben
5. reverse fault

When you complete this part, turn off the *Sideling Hill* placemark.

Instructions for Exploration 14.4 Folding and Faulting Part B:

1. Turn on and double-click the *Compression* placemark.

2. Examine the structure of the rock strata in the area around this placemark.

Exploration 14.4 Folding and Faulting Part B

B. Which of the following terms is the best geologic description for the feature seen in the vicinity of the *Compression* placemark?

1. synclinal valley
2. anticlinal valley
3. synclinal ridge
4. overturned fold
5. horst and graben

IMPORTANT: When you complete this part, turn off the Historical Imagery time slider.

Instructions for Exploration 14.4 Folding and Faulting Part C:

1. Turn on and double-click the *Offset Streams* placemark.

2. Examine the streams in this location, noting that the marked stream in this location makes dramatic 90 degree turns in its path as the result of local fault action.

Exploration 14.4 Folding and Faulting Part C

C. What type of fault is responsible for the stream pattern found in the **Offset Streams** folder?

1. reverse
2. normal
3. compression
4. strike-slip
5. expansion

Instructions for Exploration 14.4 Folding and Faulting Part D:

1. Turn on and double-click the *Oklahoma Earthquake* placemark.

2. The map shown here displays the reported shaking intensity during an earthquake that occurred on November 6, 2011.

3. Examine the evidence in this shake map, noting the location of areas of strongest shaking.

Exploration 14.4 Folding and Faulting Part D

D. Based on the evidence seen in the *Oklahoma Earthquake* layer, which of the following locations would be closest to the earthquake's epicenter?

1. Stillwater
2. Prague
3. Oklahoma City
4. Shawnee
5. Tulsa

When you complete this exploration, turn off and collapse the **14.4 Folding and Faulting** folder.

Exploration 14.4 – Folding and Faulting
Short Answer

Instructions for all Parts

1. Make sure you have opened the **KMZ** file from www.mygeoscienceplace.com.

2. From the Places panel, expand **14. Internal Processes.kmz** and then open the **14.4 Folding and Faulting** folder.

Instructions for Exploration 14.4 Folding and Faulting Short Answer A:

1. Turn on and double-click the *Dinosaur Ridge* placemark. This location shows a feature known as a hogback, specifically the Dakota Hogback.

2. Examine the relationship of this feature to the surrounding landscape, then research the term hogback as it relates to geology.

Exploration 14.4 Folding and Faulting Short Answer A

A. Describe the geologic structure of the feature seen at the *Dinosaur Ridge* placemark. Be sure to indicate how the structure of this feature relates to the geology of the surrounding region.

_____

_____

_____

_____

_____

_____

_____

Instructions for Exploration 14.4 Folding and Faulting Short Answer B:

1. Turn on the *Hayward fault photos* folder, and then double-click the Google Earth™ icon associated with that folder. This view of the San Francisco Bay area includes photographs of selected locations along the Hayward fault.

2. Click on any of the camera icons and examine images of ground impacts of this fault.

Exploration 14.4 Folding and Faulting Short Answer B

B. Use the scenes displayed in the photographs contained in the **Hayward fault photos** folder and evidence from outside sources to indicate the types of faulting that exist on the Hayward fault. Describe the impacts this fault has on the anthropogenic landscape of the San Francisco Bay metropolitan area.

_____

_____

_____

_____

_____

_____

_____

When you complete this exploration, turn off and collapse the **14.4 Folding and Faulting** folder.

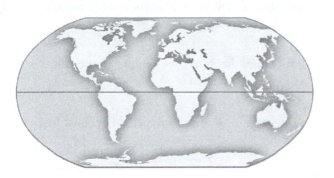

Encounter Physical Geography

Name:_____

Date: _____

Chapter 15
Weathering and Mass Wasting

Exploration 15.1 – Mechanical Weathering
Multiple Choice

The first process in the denudation of the continental surfaces is weathering. This involves breaking down rock into successively smaller components by both atmospheric and biotic processes. Examples of mechanical weathering include frost wedging and exfoliation.

Instructions for all Parts:

1. Make sure you have opened the **KMZ** file from www.mygeoscienceplace.com.

2. From the Places panel, expand **15. Weathering and Mass Wasting**.kmz and then open the **15.1 Mechanical Weathering** folder.

Instructions for Exploration 15.1 Mechanical Weathering Part A:

1. Turn on and double-click the *Openings for Weathering* placemark.

2. Examine the landscape at this location, noting evidence of potential surface openings in the local rock structure.

Exploration 15.1 Mechanical Weathering Part A

A. Which of the following surface openings are best represented by the parallel features seen at the *Openings for Weathering* placemark?

1. joints
2. faults
3. microscopic openings
4. lava vesicles
5. solution cavities

When you complete this part, turn off the *Openings for Weathering* placemark.

Instructions for Exploration 15.1 Mechanical Weathering Part B:

1. Turn on and double-click the *Half Dome* Gigapxl to fly into a portion of Yosemite National Park.

2. Within the high-resolution image, zoom to Half Dome, the dominant feature in the center right of the scene.

3. Examine the details on the surface of Half Dome, noting any evidence of mechanical weathering.

Exploration 15.1 Mechanical Weathering Part B

B. Which of the following processes of mechanical weathering is most likely responsible for the appearance of Half Dome?

1. frost wedging
2. salt wedging
3. oxidation
4. exfoliation
5. biological

When you complete this part, turn off the *Half Dome* Gigapxl image.

Instructions for Exploration 15.1 Mechanical Weathering Part C:

1. Turn on and double-click the *Cathedral Lake, Colorado* Gigapan image to fly into part of the Central Rocky Mountains.

2. Examine the details of the scene, paying particular attention to the presence of large broken rocks in the foreground.

Exploration 15.1 Mechanical Weathering Part C

C. Which of the following processes is most likely responsible for the broken rocks found in the *Cathedral Lake, Colorado* image?

1. frost wedging
2. salt wedging
3. oxidation
4. exfoliation
5. biological

When you complete this part, turn off the *Cathedral Lake, Colorado* image.

Instructions for Exploration 15.1 Mechanical Weathering Part D:

1. Turn on and open the **Exfoliation Dome** folder.

2. Examine the five placemarked locations labeled *A* through *E*, noting any potential evidence of exfoliation domes at each site.

Exploration 15.1 Mechanical Weathering Part D

D. Which of the placemarks in the **Exfoliation Dome** folder is the best example of an exfoliation dome?

1. A
2. B
3. C
4. D
5. E

When you complete this exploration, turn off and collapse the **15.1 Mechanical Weathering** folder.

Exploration 15.1 – Mechanical Weathering
Short Answer

Instructions for Exploration 15.1 Mechanical Weathering Short Answer A:

1. Make sure you have opened the **KMZ** file from www.mygeoscienceplace.com.

2. From the Places panel, expand **15. Weathering and Mass Wasting.kmz** and open the **15.1 Mechanical Weathering** folder, then turn on and double-click the *Angel's Window, Grand Canyon* Gigapxl image to fly into this scene in Grand Canyon National Park.

3. Examine the details of the rocks in the high-resolution image, noting evidence of mechanical weathering and differential weathering present in this scene.

Exploration 15.1 Mechanical Weathering Short Answer A

A. Based on the evidence in the *Angel's Window, Grand Canyon* image discuss the processes of mechanical weathering that is occurring in the Grand Canyon.

_____

_____

_____

_____

_____

_____

_____

When you complete this part, turn off the *Angel's Window, Grand Canyon* image.

Instructions for Exploration 15.1 Mechanical Weathering Short Answer B:

1. Use Google Earth™ to navigate to a location near you where mechanical weathering is occurring.

2. Examine the evidence for mechanical weathering at this location and note the potential for continued weathering action.

Exploration 15.1 Mechanical Weathering Short Answer B

B. Identify a location near you where mechanical weathering is evident. Describe the processes involved in the creation of this site. Be sure to indicate the location using latitude and longitude.

_____

_____

_____

_____

_____

_____

_____

When you complete this exploration, turn off and collapse the **15.1 Mechanical Weathering** folder.

Exploration 15.2 – Chemical and Biological Weathering
Multiple Choice

Beyond mechanical processes, rock can be broken down by chemical alteration of its minerals and/or the actions of living organisms. The degree of this chemical and biological weathering is significantly impacted by factors such as latitude and climate.

Instructions for all Parts:

1. Make sure you have opened the **KMZ** file from www.mygeoscienceplace.com.

2. From the Places panel, expand **15. Weathering and Mass Wasting.kmz** and then open the **15.2 Chemical and Biological Weathering** folder.

Instructions for Exploration 15.2 Chemical and Biological Weathering Part A:

1. Turn on and double-click the *Newspaper Rock* Gigapxl image to fly into a high-resolution photo of this famous site in eastern Utah.

2. Examine the details of the image, taking time to consider how chemical weathering played a role in its significance.

Exploration 15.2 Chemical and Biological Weathering Part A

A. Which of the following statements provides the best description of the chemical weathering seen in the *Newspaper Rock* photo?

1. The rock face has desert varnish.
2. Petroglyphs have been carved on the rock face.
3. Grasses are growing at the foot of the rock due to chemical weathering.
4. Vertical jointing is visible along the edges of the rock face.
5. Frost wedging has created the talus slope beneath the rock.

When you complete this part, turn off the *Newspaper Rock* image.

Instructions for Exploration 15.2 Chemical and Biological Weathering Part B:

1. Turn on and double-click the *Snake River* Gigapan image to fly into a high-resolution photo of this river in eastern Washington.

2. Examine the details of the image, noting the presence of biological weathering at this location.

Exploration 15.2 Chemical and Biological Weathering Part B

B. Which of the following is **not** an example of biological weathering evident in the *Snake River* image?

1. The lichen growing on the rock faces.
2. The grasses growing in some of the rock joints.
3. The trees growing in some of the joints.
4. The water interacting with the minerals in the rock.
5. Rodents utilizing joints for food storage or shelter.

When you complete this part, turn off the *Snake River* image.

Instructions for Exploration 15.2 Chemical and Biological Weathering Part C:

1. Turn on and double-click the *Lassen NP* Gigapxl image to fly into a high-resolution photo of a landscape in this Northern California volcanic park.

2. Examine the details of the image, noting the presence of biological weathering at this location.

Exploration 15.2 Chemical and Biological Weathering Part C

C. Which of the following is the best example of biological weathering evident in the *Lassen NP* image?

1. The trees that are wedging their roots into joints in the rocks.
2. The red color of the rocks as a result of oxidation.
3. Rock fragments in talus slopes as a result of frost wedging.
4. The dissolution of minerals from rocks by carbonation.
5. Crumbling rock surfaces produced by hydrolysis.

When you complete this part, turn off the *Lassen NP* image.

Instructions for Exploration 15.2 Chemical and Biological Weathering Part D:

1. Turn on and double-click the *Statue of Liberty* placemark. From the *Primary Database*, turn on the *3D Buildings* layer.

2. Examine the 3D model of the Statue of Liberty, being sure to consider how weathering may have contributed to the look of this famous landmark.

Exploration 15.2 Chemical and Biological Weathering Part D

D. Which of the following processes would best explain the green patina seen on the Statue of Liberty?

1. hydrolysis
2. carbonation
3. exfoliation
4. oxidation
5. salt wedging

IMPORTANT: Be sure to turn off the *3D Buildings* layer from the *Primary Database*.

When you complete this exploration, turn off and collapse the **15.2 Chemical and Biological Weathering** folder.

Exploration 15.2 – Chemical and Biological Weathering
Short Answer

Instructions for all Parts:

1. Make sure you have opened the **KMZ** file from www.mygeoscienceplace.com.

2. From the Places panel, expand **15. Weathering and Mass Wasting.kmz** and then open the **15.2 Chemical and Biological Weathering** folder.

Instructions for Exploration 15.2 Chemical and Biological Weathering Short Answer A:

1. Turn on and double-click the *Bayon Temple* Gigapan image to fly into a high-resolution photo of ancient temples in Southeast Asia.

2. Examine the details of the image, noting the presence of chemical weathering at this location.

Exploration 15.2 Chemical and Biological Weathering Short Answer A

A. Based on the evidence seen in the *Bayon Temple* image, provide a description of chemical weathering present at the site. Be sure to include an assessment of the local climate in your response.

_____
_____
_____
_____
_____
_____
_____

When you complete this part, turn off the *Bayon Temple* Gigapxl image.

Instructions for Exploration 15.2 Chemical and Biological Weathering Short Answer B:

1. Turn on and double-click the *East and West Mittens* Gigapan image to fly into a high-resolution photo of this iconic landscape of the American West.

2. Examine the details of the image, noting the presence of various weathering processes at this location.

Exploration 15.2 Chemical and Biological Weathering Short Answer B

B. Based on evidence found in the image, identify and describe at least two processes of chemical, biological, or mechanical weathering present at the East and West Mittens.

_____
_____
_____
_____
_____
_____

When you complete this exploration, turn off and collapse the **15.2 Chemical and Biological Weathering** folder.

Exploration 15.3 – Mass Wasting Factors
Multiple Choice

Weathered material is transported downslope through the process known as mass wasting. This movement is facilitated by factors such as slope, moisture, and underlying material. The status of these factors may combine to create events that are very slow or incredibly rapid.

Instructions for all Parts:

1. Make sure you have opened the **KMZ** file from www.mygeoscienceplace.com.

2. From the Places panel, expand **15. Weathering and Mass Wasting.kmz** and then open the **15.3 Mass Wasting Factors** folder.

Instructions for Exploration 15.3 Mass Wasting Factors Part A:

1. Turn on and double-click the *Landslide* image layer. The brown area in the center of the image is a landslide location.

2. Examine the area impacted by this landslide, being sure to note any potential causes for slope failure.

Exploration 15.3 Mass Wasting Factors Part A

A. Based on the information gathered in the *Landslide* image layer, which of the following factors is the most likely contributor to slope failure at this location?

1. The landslide was caused by a period of heavy rain.
2. Deforestation at the site of the landslide may have contributed to decreased slope stability.
3. The landslide was caused by a volcanic eruption.
4. Increased load stress due to expanding urban development resulted in a slope failure at this site.
5. The failure of an earthen dam precipitated this landslide event.

When you complete this part, turn off the *Landslide* layer.

Instructions for Exploration 15.3 Mass Wasting Factors Part B:

1. Turn on and double-click the *Slide* image layer. The dark feature running through the center part of the image represents a particular type of slide.

2. Use the navigation capabilities of Google Earth™ to explore the landscape around this site to determine the origin and type of slide shown.

Exploration 15.3 Mass Wasting Factors Part B

B. Which of the following types of slides is shown in the *Slide* image layer?

1. landslide
2. rock avalanche
3. slump
4. lahar
5. creep

When you complete this part, turn off the *Slide* image layer.

Instructions for Exploration 15.3 Mass Wasting Factors Part C:

1. Turn on and double-click the *Vajont Dam* placemark, then turn on the *3D Buildings* layer from the *Primary Database*. You should note that this dam appears to be unused, as it is not holding back a notable reservoir.

2. Use the navigational capabilities of Google Earth™ to examine the landscape around this dam, paying particular attention to any possible reason for the presence of this unused dam.

Exploration 15.3 Mass Wasting Factors Part C

C. Based on the evidence seen at the *Vajont Dam* placemark, which of the following statements provides the best explanation for the presence of an unused dam?

1. No inhabitants are in the region to benefit from hydroelectric power generation.
2. A series of landslides created massive holes in the dam structure.
3. The area of the former reservoir was filled by a series of landslides.
4. The strata of the region is too permeable to maintain surface water in large amounts.
5. The dam was just completed and the reservoir has not yet filled.

IMPORTANT: Be sure to turn off the *3D Buildings* layer from the *Primary Database*.

When you complete this part, turn off the *Vajont Dam* placemark.

Instructions for Exploration 15.3 Mass Wasting Factors Part D:

1. Turn on and double-click the *Madison River* placemark.

2. Examine this location where an earthquake-generated landslide blocked the Madison River.

Exploration 15.3 Mass Wasting Factors Part D

D. Which of the following statements is most strongly supported by your analysis of the landscape around the *Madison River* placemark?

1. The slide that dammed the Madison River originated from the slopes along the south bank of the river.
2. The shoreline of Earthquake Lake upriver from the slide site suggests that the lake is still filling.
3. The river has successfully cut through the landslide and is draining to the east.
4. The landslide-produced dam has backed up water for more than 30 km along the Madison River.
5. The earthquake that created Earthquake Lake occurred in the year 2000.

When you complete this exploration, turn off and collapse the **15.3 Mass Wasting Factors** folder.

Exploration 15.3 – Mass Wasting Factors
Short Answer

Instructions for all Parts:

1. Make sure you have opened the **KMZ** file from www.mygeoscienceplace.com.

2. From the Places panel, expand **15. Weathering and Mass Wasting.kmz** and then open the **15.3 Mass Wasting Factors** folder.

Instructions for Exploration 15.3 Mass Wasting Factors Short Answer A:

1. Double-click the *Novo Friburgo* placemark.

2. Examine the landscape in this region about 130 km northeast of Rio de Janeiro, Brazil, paying particular attention to any evidence of mass wasting.

3. Note the date indicated by the Historical Imagery slider and then do some outside research to determine the causes and impacts of any detected slides.

Exploration 15.3 Mass Wasting Factors Short Answer A

A. Provide an explanation for the large-scale mass wasting that occurred in the area around the *Novo Friburgo* placemark. Be sure to note the impacts these events had and the causes that led to their occurrence.

_____

_____

_____

_____

_____

_____

Instructions for Exploration 15.3 Mass Wasting Factors Short Answer B:

1. Double-click the *Ajka, Hungary* placemark to fly to a region in western Hungary that was impacted by a mass wasting event.

2. Examine the areas with red-tinted sediment, which is a remnant of the event. Navigate in the Google Earth™ 3D Viewer to locate the source of the sediment then use the Historical Imagery slider to determine the timeline for this event. Use any outside sources to verify the details of the event.

Exploration 15.3 Mass Wasting Factors Short Answer B

B. Based on the imagery seen in Google Earth™ and any outside research you've done, what was the source of the red-tinted sediment found at the *Ajka, Hungary* placemark? Note the latitude and longitude of the source area and provide a summary of the impacts of the event that caused this.

_____

_____

_____

_____

_____

_____

_____

When you complete this exploration, turn off and collapse the **15.3 Mass Wasting Factors** folder.

Exploration 15.4 – Landforms
Multiple Choice

The process of mass wasting creates a distinctive class of landforms that reflect the moisture content and the speed at which the process occurred. Mass wasting events can range from the relatively benign soil creep to the destructive and quick-moving landslide, each with a distinct impact on physical landscapes.

Instructions for all Parts:

1. Make sure you have opened the **KMZ** file from www.mygeoscienceplace.com.

2. From the Places panel, expand **15. Weathering and Mass Wasting.kmz** and then open the **15.4 Mass Wasting Landforms** folder.

3. The five placemarks labeled *A* through *E* represent examples of landscapes affected by mass wasting events.

4. Navigate to each of the five placemarks and evaluate the evidence in the landscape to determine the type of mass wasting that has occurred at each site.

Exploration 15.4 Mass Wasting Landforms Part A

A. Which of the placemarks in the **Mass Wasting Landforms** folder is the best example of a rock glacier?

1. A
2. B
3. C
4. D
5. E

Exploration 15.4 Mass Wasting Landforms Part B

B. Which of the placemarks in the **Mass Wasting Landforms** folder is the best example of a talus slope?

1. A
2. B
3. C
4. D
5. E

Exploration 15.4 Mass Wasting Landforms Part C

C. Which of the placemarks in the **Mass Wasting Landforms** folder is the best example of an earthflow?

1. A
2. B
3. C
4. D
5. E

Exploration 15.4 Mass Wasting Landforms Part D

D. Which of the placemarks in the **Mass Wasting Landforms** folder is the best example of creep?

1. A
2. B
3. C
4. D
5. E

When you complete this exploration, turn off and collapse the **15.4 Mass Wasting Landforms** folder.

Exploration 15.4 – Landforms
Short Answer

Instructions for all Parts:

1. Make sure you have opened the **KMZ** file from www.mygeoscienceplace.com.

2. From the Places panel, expand **15. Weathering and Mass Wasting.kmz** and then open the **15.4 Mass Wasting Landforms** folder.

Instructions for Exploration 15.4 Mass Wasting Landforms Short Answer A:

1. Double-click the *Silverton* placemark to fly to this town in southwest Colorado.

2. Examine the landscape surrounding Silverton, being sure to note any evidence of mass wasting events that may have occurred here.

3. Assess the potential risks mass wasting events may have on the town of Silverton and its residents.

Exploration 15.4 Mass Wasting Landforms Short Answer A

A. Describe the evidence of past mass wasting events seen in at the *Silverton* placemark. What impacts might mass wasting events have on the town of Silverton and the surrounding landscape?

_____

_____

_____

_____

_____

_____

_____

Instructions for Exploration 15.4 Mass Wasting Landforms Short Answer B:

1. Double-click the *Site A* placemark to fly to a landscape in central Ireland.

2. Evaluate the landscape, paying particular attention to the possibility for and type of mass wasting that might occur in this location. You should consider landscape characteristics such as slope, vegetation, and climate.

3. Double-click the *Site B* placemark to fly to a landscape in the North African nation of Morocco.

4. Repeat the same evaluation as at *Site A*, being sure to note potential for mass wasting at this location.

Exploration 15.4 Mass Wasting Landforms Short Answer B

B. Based on factors of slope, vegetation, and climate, describe a specific type of mass wasting that you would expect to see at *Site A* and *Site B*. Why would mass wasting processes be different between these two sites?

_____

_____

_____

_____

_____

_____

_____

When you complete this exploration, turn off and collapse the **15.4 Mass Wasting Landforms** folder.

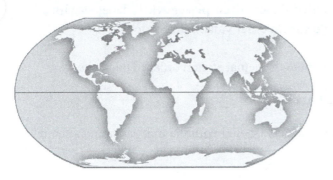

Encounter Physical Geography

Name:_____

Date: _____

**Chapter 16
Fluvial Processes**

**Exploration 16.1 – Streams and Stream Systems
Multiple Choice**

Streams and stream valleys provide some of the most conspicuous natural patterns on the landscape. These patterns relate to the relative magnitude of the stream as well as the surface upon which they flow.

Instructions for all Parts

1. Make sure you have opened the **KMZ** file from www.mygeoscienceplace.com.

2. From the Places panel, expand **16. Fluvial Processes**.kmz and then open the **16.1 Streams and Stream Systems** folder.

Instructions for Exploration 16.1 Streams and Stream Systems Part A:

1. Turn on and double-click the **Drainage Divide** folder. The five lines labeled *A* through *E* represent potential locations of the boundary between local drainage basins.

2. Examine the five lines, noting the likelihood of a drainage divide occurring at each location.

Exploration 16.1 Streams and Stream Systems Part A

A. Which line contained in the **Drainage Divide** folder best represents a drainage divide?

1. A
2. B
3. C
4. D
5. E

When you complete this part, turn off and collapse the **Drainage Divide** folder.

Instructions for Exploration 16.1 Streams and Stream Systems Part B:

1. Turn on and double-click the **Stream Order** folder. Each placemark is labeled with a number from 1 to 5, representing a <u>possible</u> stream order value for each location.

2. Examine the five placemarks, being sure to note the <u>actual</u> stream order of each site.

Exploration 16.1 Streams and Stream Systems Part B:

B. Which of the placemarks in the **Stream Order** folder is labeled with the most accurate stream order number?

1. 1
2. 2
3. 3
4. 4
5. 5

When you complete this part, turn off and collapse the **Stream Order** folder.

Instructions for Exploration 16.1 Streams and Stream Systems Part C:

1. Turn on and open the **Stream Drainage Patterns** folder. The five placemarks in this folder represent examples of different stream drainage patterns around the world.

2. Examine the landscapes at each of the placemarks, noting the drainage pattern that exists at each location.

Exploration 16.1 Streams and Stream Systems Part C

C. Which of the placemarks in the **Stream Drainage Patterns** folder represents a radial drainage pattern?

1. A
2. B
3. C
4. D
5. E

When you complete this part, turn off and collapse the **Stream Drainage Patterns** folder.

Instructions for Exploration 16.1 Streams and Stream Systems Part D:

1. Turn on and open the **Stream Channel Patterns** folder. The five placemarks labeled *A* through *E* show a selection of stream channels in a variety of places around the world.

2. Examine each of the placemarked locations, paying particular attention to the stream channel patterns that exist at each location.

Exploration 16.1 Streams and Stream Systems Part D

D. Which of the placemarks in the **Stream Channel Patterns** folder best resembles a braided stream pattern?

1. A
2. B
3. C
4. D
5. E

When you complete this exploration, turn off and collapse the **16.1 Streams and Stream Systems** folder.

Exploration 16.1 – Streams and Stream Systems
Short Answer

Instructions for All Short Answer

1. Make sure you have opened the **KMZ** file from www.mygeoscienceplace.com.

2. From the Places panel, expand **16. Fluvial Processes.kmz** and then open the **16.1 Streams and Stream Systems** folder.

Instructions for Exploration 16.1 – Streams and Stream Systems Short Answer A:

1. Turn on and double-click the **Stream Order** folder. The placemarked streams in this view all eventually flow into the Columbia River and are part of the larger Columbia River drainage basin.

2. Zoom out to examine the extent of the Columbia's watershed. Be sure to note the physical barriers that would be the likely drainage divides between the Columbia and other rivers.

Exploration 16.1 – Streams and Stream Systems Short Answer A

A. Utilizing Google Earth™ and any additional outside resources, identify the states and/or provinces that are included in the Columbia River watershed/drainage basin.

_____

_____

_____

_____

_____

_____

_____

When you complete this part, turn off and collapse the **Stream Order** folder.

Instructions for Exploration 16.1 – Streams and Stream Systems Short Answer B:

1. Navigate to your location in Google Earth™.

2. Examine the dominant stream drainage pattern that exists in your area. Be sure to consider where streams or any surface runoff would flow.

Exploration 16.1 – Streams and Stream Systems Short Answer B

B. Identify your location in latitude and longitude. What stream drainage pattern is most apparent in your area?

_____

_____

_____

_____

_____

_____

_____

When you complete this exploration, turn off and collapse the **16.1 Streams and Stream Systems** folder.

Exploration 16.2 – Fluvial Erosion and Deposition
Multiple Choice

As water moves across the surface of the planet it has a remarkable capacity to erode and transport sediment. Factors such as water volume, slope, and streambed composition affect rates of erosion and deposition. Stream erosion and deposition can create a series of dynamic floodplain landforms.

Instructions for all Parts:

1. Make sure you have opened the **KMZ** file from www.mygeoscienceplace.com.

2. From the Places panel, expand **16. Fluvial Processes.kmz** and then open the **16.2 Fluvial Erosion and Deposition** folder.

Instructions for Exploration 16.2 Fluvial Erosion and Deposition Part A:

1. Turn on and double-click the **Stream Gradient** folder. *Stream Gradient Point A* and *Stream Gradient Point B* represent the upstream and downstream extents of a portion of the Arkansas River in central Colorado.

2. Utilize the measuring and elevation capabilities of Google Earth™ to determine the correct stream gradient between the two placemarks.

Exploration 16.2 Fluvial Erosion and Deposition Part A

A. What is the stream gradient of the Arkansas River between the two placemarks in the **Stream Gradient** folder?

1. 38.43 meters/kilometer
2. 11.64 meters/kilometer
3. 13.09 meters/kilometer
4. 23.83 meters/kilometer
5. 7.22 meters/kilometer

When you complete this part, turn off and collapse the **Stream Gradient** folder.

Instructions for Exploration 16.2 Fluvial Erosion and Deposition Part B:

1. Turn on and open the **Stream Types** folder. The five placemarks in this folder represent examples of different stream types from around the world.

2. Evaluate the stream identified by each placemark, noting the status of the stream type at each location.

Exploration 16.2 Fluvial Erosion and Deposition Part B

B. Which of the placemarks in the **Stream Types** folder represents an ephemeral stream?

1. A
2. B
3. C
4. D
5. E

When you complete this part, turn off and collapse the **Stream Types** folder.

Instructions for Exploration 16.2 Fluvial Erosion and Deposition Part C:

1. Turn on and double-click the **Floodplain Landforms** folder.

2. Examine the floodplain features identified by the five placemarks, being sure to consider relative rates of active erosion or deposition at each site.

Exploration 16.2 Fluvial Erosion and Deposition Part C

C. Which of the labeled placemarks in the **Floodplain Landforms** folder would have the highest active rate of deposition?

1. A
2. B
3. C
4. D
5. E

When you complete this part, turn off and collapse the **Floodplain Landforms** folder.

Instructions for Exploration 16.2 Fluvial Erosion and Deposition Part D:

1. Turn on and open the **Davis' Geomorphic Cycle** folder. Each placemark labeled *A* through *E* represents fluvial landscapes that could be classified within Davis' Geomorphic Cycle model.

2. View the placemarked locations, evaluating each for its most likely designation in Davis' Geomorphic Cycle.

Exploration 16.2 Fluvial Erosion and Deposition Part D

D. Which of the following placemarked locations in the **Davis' Geomorphic Cycle** folder best illustrates Davis' idea of an old age landscape?

1. A
2. B
3. C
4. D
5. E

When you complete this exploration, turn off and collapse the **16.2 Fluvial Erosion and Deposition** folder.

Exploration 16.2 Fluvial Erosion and Deposition
Short Answer

Instructions for all Parts:

1. Make sure you have opened the **KMZ** file from www.mygeoscienceplace.com.

2. From the Places panel, expand **16. Fluvial Processes.kmz** and then open the **16.2 Fluvial Erosion and Deposition** folder.

Instructions for Exploration 16.2 Fluvial Erosion and Deposition Short Answer A:

1. Double-click the **Floodplain Landforms** folder.

2. Examine the floodplain landscape along this portion of the Mississippi River, being sure to identify four typical floodplain features.

Exploration 16.2 Fluvial Erosion and Deposition Short Answer A

A. Based on the evidence in the landscape shown in the **Floodplain Features** folder, identify four typical floodplain features. Be sure to identify each by their appropriate name and latitude and longitude.

_____
_____
_____
_____
_____
_____
_____

When you complete this part, collapse the **Floodplain Landforms** folder.

Instructions for Exploration 16.2 – Fluvial Erosion and Deposition Short Answer B:

1. Navigate to your location in Google Earth™. Examine the surrounding region and identify the highest order stream in your area.

2. Calculate the stream gradient over a 10 km stretch of this stream. Compare this stream gradient to the gradient found between the placemarks in the **Stream Gradient** folder.

3. Evaluate the landscape created by the highest order stream in your area in the context of Davis' Geomorphic Cycle.

Exploration 16.2 – Fluvial Erosion and Deposition Short Answer B

B. What is the highest order stream in your area? Identify the name of that stream and your location. What is its stream gradient? How different is this gradient from that of the stream in the **Stream Gradient** folder? What stage of Davis' Geomorphic Cycle would best describe the stream landscape of this location?

_____
_____
_____
_____
_____
_____
_____

When you complete this exploration, turn off and collapse the **16.2 Fluvial Erosion and Deposition** folder.

Exploration 16.3 – Shaping Valleys
Multiple Choice

Most flowing water is contained by valleys. These valleys can be deepened, widened, and lengthened as a stream cuts down to its base level.

Instructions for all Parts:

1. Make sure you have opened the **KMZ** file from www.mygeoscienceplace.com.

2. From the Places panel, expand **16. Fluvial Processes.kmz** and then open the **16.3 Shaping Valleys** folder.

Instructions for Exploration 16.3 Shaping Valleys Part A:

1. Turn on and double-click the **Stream Capture** folder to fly to an area near the mouth of the Murray River in South Australia. The Murray River currently empties into the Great Australian Bight 70 km south of Adelaide. The current mouth developed after the original channel was cut off by tectonic uplift.

2. Examine the five lines in the **Stream Capture** folder, noting evidence of a former stream channel along each path.

Exploration 16.3 Shaping Valleys Part A

A. Which of the generalized channel paths seen in the **Stream Capture** folder most likely represents the previous outlet channel of the Murray River?

1. A
2. B
3. C
4. D
5. E

When you complete this part, turn off and collapse the **Stream Capture** folder.

Instructions for Exploration 16.3 Shaping Valleys Part B:

1. Turn on and double-click the *Knickpoint* placemark to fly to Niagara Falls.

2. Examine the Niagara Falls area, being sure to utilize the tilt capabilities of Google Earth™ to get a better understanding of this knickpoint.

Exploration 16.3 Shaping Valleys Part B

B. Which of the following statements is best supported by the evidence found at the *Knickpoint* placemark?

1. The bridge to the north of the falls is in danger from knickpoint migration.
2. Bird Island and Robinson Island will likely grow because of deposition in the near future.
3. The process of knickpoint migration will ensure that the more spectacular Horseshoe Falls will remain on the US side of the US–Canadian border.
4. The Niagara Escarpment is being eroded to the south and east by the Niagara River in the vicinity of Niagara Falls.
5. Tourist boat excursions approach the base of the falls from an upstream direction.

When you complete this part, turn off the *Knickpoint* placemark.

Instructions for Exploration 16.3 Shaping Valleys Part C:

1. Turn on and open the **Stream Load** folder. Each of the five placemarks labeled *A* through *E* shows a confluence point between two rivers.

2. Examine each of the placemarked confluence sites, noting the apparent difference in suspended load between the two merging streams.

Exploration 16.3 Shaping Valleys Part C

C. Which of the confluence sites placemarked in the **Stream Load** folder best displays a difference in suspended load?

1. A
2. B
3. C
4. D
5. E

When you complete this part, turn off and collapse the **Stream Load** folder.

Instructions for Exploration 16.3 Shaping Valleys Part D:

1. Turn on and open the **Deltas** folder. The five placemarks labeled *A* through *E* mark five rivers throughout the world.

2. Fly to each placemark and navigate downstream to the mouth of each respective river. Deltas form based on the relative amount of sediment carried by a river to its mouth and the presence or absence of ocean currents and wave action at the coast. Consider these factors as you note the presence of and relative size of the delta at the mouth of each river.

Exploration 16.3 Shaping Valleys Part D

D. Which of the placemarks in the **Deltas** folder is located on the river that flows to the largest delta, as defined by the width of the largest part of the delta at the mouth of the river?

1. A
2. B
3. C
4. D
5. E

When you complete this exploration, turn off and collapse the **16.3 Shaping Valleys** folder.

Exploration 16.3 – Shaping Valleys
Short Answer

Instructions for all Parts:

1. Make sure you have opened the **KMZ** file from www.mygeoscienceplace.com.

2. From the Places panel, expand **16. Fluvial Processes.kmz** and then open the **16.3 Shaping Valleys** folder.

Instructions for Exploration 16.3 Shaping Valleys Short Answer A:

1. Turn on and double-click the *Structure* placemark. At this location, the river seems to defy its surrounding terrain.

2. Examine the terrain shown at this placemark and then use your textbook to research the possible causes for this landscape.

Exploration 16.3 Shaping Valleys Short Answer A

A. Based on the evidence seen at the *Structure* placemark and any outside sources, identify the process that allowed this stream to flow through the mountain. Be sure to use appropriate terminology in your response.

_____

_____

_____

_____

_____

_____

_____

When you complete this question, turn off the *Structure* placemark.

Instructions for Exploration 16.3 Shaping Valleys Short Answer B:

1. Turn on and double-click the *Headwaters* placemark to fly to the headwaters of a major American river. In theory, if you emptied a glass of water into this stream it would follow its entire course to the ocean.

2. Use the navigation capabilities in Google Earth™ to trace the route of this stream to the ocean, noting the names of major rivers or water bodies that are encountered along the way.

Exploration 16.3 – Shaping Valleys Short Answer B

B. What named rivers and water bodies are encountered as this stream flows to the ocean? Through what states does the route pass?

_____

_____

_____

_____

_____

_____

_____

When you complete this exploration, turn off and collapse the **16.3 Shaping Valleys** folder.

Exploration 16.4 – Floodplains
Multiple Choice

Floodplains have historically been important locations of human settlement. Agriculture, urban development, and transportation routes are suited to these flat locations. However, most streams are prone to occasional flooding and thus present a potential hazard for the human elements of the landscape that exist in these areas.

Instructions for all Parts

1. Make sure you have opened the **KMZ** file from www.mygeoscienceplace.com.

2. From the Places panel, expand **16. Fluvial Processes.kmz** and then open the **16.4 Floodplains** folder.

Instructions for Exploration 16.4 Floodplains Part A:

1. Turn on and double-click the **Cedar Rapids** folder. Cedar Rapids sits along the banks of the Cedar River in eastern Iowa. In the area of downtown Cedar Rapids, the 100-year floodplain covers all areas that are up to approximately 9 meters above the normal river level.

2. Evaluate the five placemarks labeled *A* through *E*, being sure to note their proximity to the 100-year floodplain for the Cedar River.

Exploration 16.4 Floodplains Part A

A. Which of the placemarks in the **Cedar Rapids** folder is located outside the 100-year floodplain zone?

1. A
2. B
3. C
4. D
5. E

When you complete this part, turn off and collapse the **Cedar Rapids** folder.

Instructions for Exploration 16.4 Floodplains Part B:

1. Turn on and double-click the **Runoff** folder. The five polygons delineate different portions of the Los Angeles metro area.

2. Double-click the polygons in the folder to fly to and examine each urban landscape and evaluate the potential for infiltration and runoff at each site.

Exploration 16.4 Floodplains Part B

B. Which of the polygons in the **Runoff** folder would likely have the greatest amount of surface runoff during a rain event?

1. A
2. B
3. C
4. D
5. E

When you complete this part, turn off and collapse the **Runoff** folder.

Instructions for Exploration 16.4 Floodplains Part C:

1. Turn on and open the **Dams** folder.

2. Double-click the *Parker Dam* placemark to fly to this dam on the Colorado River that impounds Lake Havasu.

3. Use the navigation controls in Google Earth™ to trace the Colorado River upstream all the way to the *Glen Canyon Dam* placemark. As you navigate the river, you should be sure to note the presence of any additional dams and their respective reservoirs. You may find it helpful to turn on the *3D Buildings* layer from the *Primary Database*, as many dams have 3D models rendered in Google Earth™.

Exploration 16.4 Floodplains Part C

C. How many dams, **not including** the two dams placemarked in the **Dams** folder, impound the Colorado River between Glen Canyon and Lake Havasu?

    1. 0
    2. 1
    3. 2
    4. 3
    5. 4

IMPORTANT: Be sure to turn off the *3D Buildings* layer from the *Primary Database* if you turned it on.

When you complete this part, turn off and collapse the **Dams** folder.

Instructions for Exploration 16.4 Floodplains Part D:

1. Turn on and open the **Water Diversion** folder. The five placemarks labeled *A* through *E* show a sample of agricultural landscapes around the globe.

2. Examine each of the placemarks, noting any modification of surface stream flow for agricultural purposes.

Exploration 16.4 Floodplains Part D

D. Which of the placemarks in the **Water Diversion** folder ***does not*** include the alteration of surface stream flow for the purpose of agriculture?

    1. A
    2. B
    3. C
    4. D
    5. E

When you complete this exploration, turn off and collapse the **16.4 Floodplains** folder.

Exploration 16.4 – Floodplains
Short Answer

Instructions for all Parts:

1. Make sure you have opened the **KMZ** file from www.mygeoscienceplace.com.

2. From the Places panel, expand **16. Fluvial Processes.kmz** and then open the **16.4 Floodplains** folder.

Instructions for Exploration 16.4 Floodplains Short Answer A:

1. Double-click the **Cedar Rapids** folder. The 100-year floodplain for the Cedar River in the Cedar Rapids area inundates areas less than 9 meters above the normal river level of 214 meters.

2. Examine the Google Earth™ imagery for Cedar Rapids, paying particular attention to the approximate location of the 100-year floodplain, as indicated by elevation in the status bar, and the types of activities located within that zone.

Exploration 16.4 Floodplains Short Answer A

A. How would Cedar Rapids, Iowa, be impacted during a 100-year flood event? Describe the impacts to commercial, industrial, and residential developments in the areas around downtown Cedar Rapids.

_____
_____
_____
_____
_____

When you complete this part, turn off and collapse the **Floodplain Uses** folder.

Instructions for Exploration 16.4 Floodplains Short Answer B:

1. Fly to your location in Google Earth™.

2. Flash flooding occurs more frequently in areas where rainwater infiltration is limited by the presence of impervious surface cover, such as paved surfaces in urban areas. Evaluate the streams in your area, taking note of the risk for flash flooding for each.

Exploration 16.4 Floodplains Short Answer B

B. Based on the evidence in Google Earth™, name a stream in your area that is likely to have a high risk for flash flooding. Describe the characteristics that contribute to this high risk for this watershed. Be certain to provide the name of your location and the latitude and longitude for at least one landscape example that provides support for your assessment.

_____
_____
_____
_____
_____

When you complete this exploration, turn off and collapse the **16.4 Floodplains** folder.

Encounter Physical Geography

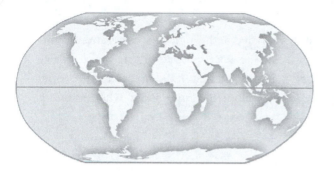

Name:_____

Date: _____

Chapter 17
Karst and Hydrothermal Processes

Exploration 17.1 – Karst Features
Multiple Choice

Water, acting as a solvent, can be very effective at dissolving certain types of underground rocks. This can create unique surface topography, such as tower karst, as well as excavated underground caverns.

Instructions for all Parts:

1. Make sure you have opened the **KMZ** file from www.mygeoscienceplace.com.

2. From the Places panel, expand **17. Karst and Hydrothermal Processes.kmz** and then open the **17.1 Karst Features** folder.

Instructions for Exploration 17.1 Karst Features Part A:

1. Turn on and double-click on the *Surface Water* placemark.

2. Evaluate the surrounding landscape in regard to the presence of surface water.

Exploration 17.1 Karst Features Part A

A. Which of the following statements is best supported by the visible evidence in the 3D Viewer window surrounding the *Surface Water* placemark?

1. The light-colored linear feature running roughly north-south is the only major stream in this area.
2. This region is arid and has very little precipitation to feed surface streams.
3. Because of the lack of surface water, no agricultural activity is present.
4. This region lacks clearly defined surface drainage due to the presence of soluble surface rocks.
5. The lack of surface water in this region means that large-scale human civilization could not flourish here.

When you complete this part, turn off the *Surface Water* placemark.

Instructions for Exploration 17.1 Karst Features Part B:

1. Turn on and open the **Tower Karst** folder.

2. Examine each of the five placemarked locations labeled *1* through *5*, paying attention to the landscape features at each location. Turn on the *Photos* layer from the *Primary Database* and open available images to get a more detailed look at each landscape. Placemark *1* utilizes Historical Imagery; be sure to close the time slider when viewing other placemarks.

Exploration 17.1 Karst Features Part B

B. Which placemarks in the **Tower Karst** folder provide the best examples of tower karst landscapes?

    1. 1 & 2
    2. 1 & 3
    3. 3 & 4
    4. 3 & 5
    5. 2 & 4

IMPORTANT: Turn off the *Photos* layer from the *Primary Database*.

When you complete this part, turn off and collapse the **Tower Karst** folder.

Instructions for Exploration 17.1 Karst Features Part C:

1. Turn on and open the **Sinkhole** folder.

2. Examine each of the placemarked locations, being sure to note the presence or absence of sinkholes in each landscape.

Exploration 17.1 Karst Features Part C

C. Which of the placemarked locations in the **Sinkhole** folder is the best example of a sinkhole?

    1. A
    2. B
    3. C
    4. D
    5. E

When you complete this part, turn off and collapse the **Sinkhole** folder.

Instructions for Exploration 17.1 Karst Features Part D:

1. Turn on and double-click the *Carlsbad Caverns* 360 Degree View image to fly into a room in New Mexico's Carlsbad Caverns National Park.

2. Explore the image of the cavern, paying particular attention to the types of speleothems present.

Exploration 17.1 Karst Features Part D

D. Which of the following statements is best supported by the *Carlsbad Caverns* 360 Degree View image and your knowledge of speleothems?

1. Numerous stalagmites are visible protruding from the cavern's ceiling.
2. This cavern has no evidence of soda straw formation.
3. No columns are evident in this cavern.
4. Numerous stalactites are visible protruding from the cavern floor.
5. This cavern is in the decoration stage.

When you complete this exploration, turn off and collapse the **17.1 Karst Features** folder.

Exploration 17.1 – Karst Features
Short Answer

Instructions for all Parts:

1. Make sure you have opened the **KMZ** file from www.mygeoscienceplace.com.

2. From the Places panel, expand **17. Karst and Hydrothermal Processes.kmz** and then open the **17.1 Karst Features** folder.

Instructions for Exploration 17.1 Karst Features Short Answer A:

1. Turn on and double-click the *Carlsbad Caverns* 360 Degree image to fly into a room in New Mexico's Carlsbad Caverns National Park.

2. Explore the image of the cavern and consider the processes required to form this landscape.

Exploration 17.1 Karst Features Short Answer A

A. Describe the likely progression of cavern formation and decoration that resulted in the features seen in the *Carlsbad Caverns* 360 Degree View image.

_____

_____

_____

_____

_____

When you complete this part, turn off the *Carlsbad Caverns* 360 Degree image.

Instructions for Exploration 17.1 Karst Features Short Answer B:

1. Double-click the *Surface Water* placemark. The region shown here was inhabited by the Mayan Civilization during parts of the Classic Period.

2. Examine the landscapes of the region, then use outside sources to determine how the Mayan people were able to survive and thrive in an area of karst topography.

Exploration 17.1 Karst Features Short Answer B

B. Based on your research about the landscape found at the *Surface Water* placemark, explain how the Mayans were able to build their civilization in a region of karst topography and limited surface water. Be sure to indicate how residents were able to access and store adequate water supplies.

_____

_____

_____

_____

_____

_____

_____

When you complete this exploration, turn off and collapse the **17.1 Karst Features** folder.

Exploration 17.2 – Hydrothermal Features
Multiple Choice

In a limited number of locations around the world, very hot sub-surface water makes its way through natural openings in the crust to emerge as steam or hot water. The resulting features, such as hot springs, geysers, and fumaroles, create distinctive landscapes where they occur.

Instructions for all Parts:

1. Make sure you have opened the **KMZ** file from www.mygeoscienceplace.com.

2. From the Places panel, expand **17. Karst and Hydrothermal Processes.kmz** and then open the **17.2 Hydrothermal Features** folder.

Instructions for Exploration 17.2 Hydrothermal Features Part A:

1. Turn on and double-click the *Mammoth Hot Springs* Gigapan image to fly into a scene in the northern part of Yellowstone National Park.

2. Examine the travertine terraces at Mammoth Hot Springs, paying particular attention to evidence of the processes that built these features.

Exploration 17.2 Hydrothermal Features Part A

A. What type of rock is likely involved in the processes that created the features in the *Mammoth Hot Springs* Gigapan image?

1. clastic sedimentary rock
2. extrusive igneous rock
3. intrusive igneous (plutonic) rock
4. carbonate sedimentary rock
5. foliated metamorphic rock

When you complete this part, turn off the *Mammoth Hot Springs* Gigapan image.

Instructions for Exploration 17.2 Hydrothermal Features Part B:

1. Turn on and double-click the *Yellowstone Feature* 360 Degree image to fly into a landscape in the central part of Yellowstone National Park.

2. Evaluate this photo and the Google Earth™ imagery at this location, noting the hydrothermal features that exist at this site.

Exploration 17.2 Hydrothermal Features Part B

B. Which of the following hydrothermal features is shown in the *Yellowstone Feature* 360 Degree image?

1. hot spring
2. geyser
3. fumarole
4. doline
5. speleothem

When you complete this part, turn off the *Yellowstone Feature* 360 Degree image.

Instructions for Exploration 17.2 Hydrothermal Features Part C:

1. Turn on and double-click the *New Zealand Hydrothermal* 360 Degree image to fly into a hydrothermal landscape near Rotorua, New Zealand.

2. Examine the landscape shown in the 360 Degree image, the Google Earth™ imagery of the area, and images from the *Photos* layer in the *Primary Database*. Be sure to consider the way this landscape was formed.

Exploration 17.2 Hydrothermal Features Part C

C. Which of the following statements is best supported by the landscape at the *New Zealand Hydrothermal* 360 Degree image?

1. Fumaroles are not present at this site.
2. This area does not appear to have hot springs.
3. This image demonstrates that hydrothermal regions are too dangerous for permanent human settlement.
4. There is no evidence of geysers in this area
5. The images associated with this site suggest that underground water is coming in contact with heated rocks or magma.

IMPORTANT: If you turned on the *Photos* layer from the *Primary Database*, be sure to turn it off.

When you complete this part, turn off the *New Zealand Hydrothermal* image.

Instructions for Exploration 17.2 Hydrothermal Features Part D:

1. Turn on and open the **Hydrothermal Region** folder.

2. Examine each of the five placemarks, noting the likelihood of encountering hydrothermal features in each location.

Exploration 17.2 Hydrothermal Features Part D

D. Which of the locations placemarked in the **Hydrothermal Region** folder is located in an area you would most expect to find hydrothermal features?

1. A
2. B
3. C
4. D
5. E

When you complete this exploration, turn off and collapse the **17.2 Hydrothermal Features** folder.

Exploration 17.2 – Hydrothermal Features
Short Answer

Instructions for all Parts:

1. Make sure you have opened the **KMZ** file from www.mygeoscienceplace.com.

2. From the Places panel, expand **17. Karst and Hydrothermal Processes.kmz** and then open the **17.2 Hydrothermal Features** folder.

Instructions for Exploration 17.2 Hydrothermal Features Short Answer A:

1. Turn on and double-click the *Yellowstone Feature* 360 Degree image to fly into a hydrothermal landscape in central Yellowstone National Park.

2. Examine the landscape surrounding this hydrothermal feature, paying particular attention to the status of vegetation on the surrounding hillsides.

Exploration 17.2 Hydrothermal Features Short Answer A

A. Describe the vegetation on the hillside to the south of the *Yellowstone Feature* 360 Degree image location. Do you think the condition of this vegetation is a result of toxic gasses emitted from the hydrothermal features at this location? Why or why not? If not, what might be an alternate explanation for the condition of this forest?

_____

_____

_____

_____

_____

_____

_____

When you complete this part, turn off the *Yellowstone Feature* 360 Degree image.

Instructions for Exploration 17.2 Hydrothermal Features Short Answer B:

1. Use Google Earth™ to locate the closest area of hydrothermal activity to your location.

2. Do some outside research to determine the geologic reasons for this activity.

Exploration 17.2 Hydrothermal Features Short Answer B

B. What is the closest area of hydrothermal activity to your location? What accounts for this hydrothermal activity? Be sure to identify your location.

_____

_____

_____

_____

_____

_____

_____

When you complete this exploration, turn off and collapse the **17.2 Hydrothermal Features** folder.

Exploration 17.3 – Geothermal Energy
Multiple Choice

Heat generated in the Earth's crust by ongoing radioactive decay of certain elements can provide an opportunity for energy development where the crust is particularly thin. Geothermal energy is one renewable energy source humans may look to in the future as a replacement for or supplement to existing energy resources.

Instructions for all Parts:

1. Make sure you have opened the **KMZ** file from www.mygeoscienceplace.com.

2. From the Places panel, expand **17. Karst and Hydrothermal Processes.kmz** and then open the **17.3 Geothermal Energy** folder.

3. The data for the following parts was created by the Southern Methodist University Geothermal Laboratory and released through the Google.org Enhanced Geothermal Systems (EGS) website.

Instructions for Exploration 17.3 Geothermal Energy Part A:

1. From the **Potential by State** folder, turn on and double-click the **3-D Extrusion** folder. States are shown in three dimensions based on their technical potential for development of enhanced geothermal systems.

2. Examine the patterns of geothermal energy potential, noting the states with the highest potential.

Exploration 17.3 Geothermal Energy Part A

A. Which states have the top three values for technical EGS potential based on data shown in the **3-D Extrusion** folder?

1. Nevada, Montana, Wyoming
2. Nevada, Idaho, Colorado
3. Oregon, Colorado, California
4. Idaho, Arizona, Utah
5. Colorado, West Virginia, Idaho

Instructions for Exploration 17.3 Geothermal Energy Part B:

1. Click on one of the states displayed in the **3-D Extrusion** folder. The data will show the geothermal energy potential if 2 percent, 14 percent, or 20 percent of existing resources are recovered or developed.

2. Click the hyperlink on the **US Data Summary** folder to view the same data for the entire United States. Evaluate the data, noting the potential for geothermal energy in the United States.

Exploration 17.3 Geothermal Energy Part B

B. Which of the following statements best reflects the data available from the information provided in the **US Data Summary** folder hyperlink?

1. With a 14 percent recovery rate, the EGS potential will not equal the installed generation capacity.
2. Alaska accounts for half of the EGS potential shown in the data.
3. The generation potential from areas with a temperature of 300°C accounts for 10 percent of all potential.
4. Most of the EGS potential occurs at a depth of 6.5 kilometers.
5. The EGS energy potential is measured in geotherms.

When you complete this part, close the pop-up window for the **US Data Summary** folder, then turn off the **3-D Extrusion** folder.

Instructions for Exploration 17.3 Geothermal Energy Part C:

1. Turn on the **Heat Flow** folder. This map shows the Estimated Surface Heat Flow (in mW/m2) for the continental United States. Click the link on the **Heat Flow** folder for more information on the measurement of surface heat flow.

2. Examine the pattern of surface heat flow across the continental United States, noting areas with the highest estimated values.

Exploration 17.3 Geothermal Energy Part C

C. Which of the following cities is located in the area of highest estimated surface heat flow based on the data in the **Heat Flow** folder?

1. San Francisco
2. Kansas City
3. Los Angeles
4. Chicago
5. Washington, DC

When you complete this part, turn off the **Heat Flow** folder.

Instructions for Exploration 17.3 Geothermal Energy Part D:

1. Open the **Temp at Depth Maps** folder. The layers in this folder show the estimated temperature at various depths below the surface.

2. Turn on double-click the *Temp at 6.5 km* layer. A depth of 6.5 kilometers is generally considered the approximate limit of current, feasible geothermal energy recovery.

3. Evaluate the sub-surface temperature distribution across the United States.

D. Exploration 17.3 Geothermal Energy Part D

D. Which of the following statements is best supported by the data shown by the *Temp at 6.5 km* layer in the **Temp at Depth Maps** folder?

1. The highest temperatures that occur in the continental United States do not exceed 250°C.
2. No areas with temperatures over 175°C lie east of the Mississippi River.
3. The largest areas with temperatures over 300°C are near Yellowstone National Park and in California's Imperial Valley.
4. The highest temperatures are found in the eastern part of the country.
5. Idaho exhibits the highest range in temperatures

When you complete this exploration, turn off and collapse the **17.3 Geothermal Energy** folder.

Exploration 17.3 – Geothermal Energy
Short Answer

Instructions for all Parts:

1. Make sure you have opened the **KMZ** file from www.mygeoscienceplace.com.

2. From the Places panel, expand **17. Karst and Hydrothermal Processes.kmz** and then open the **17.3 Geothermal Energy** folder.

Instructions for Exploration 17.3 Geothermal Energy Short Answer A:

1. From the **Potential by State** folder, turn on the **3-D Extrusion** folder.

2. Click on your state to examine its Enhanced Geothermal Systems potential. Be sure to compare the recovery estimates with the installed generation capacity that already exists.

Exploration 17.3 Geothermal Energy Short Answer A

A. How many megawatts (MW) of EGS potential does your state possess with 14 percent recovery? With 20 percent recovery? How does this compare with the current installed generation capacity of all sources? How might geothermal energy impact total energy production in your state? Be sure to indicate your state in your response.

_____

When you complete this part, turn off the **3-D Extrusions** folder and close any open pop-up windows.

Instructions for Exploration 17.3 Geothermal Energy Short Answer B:

1. Open and turn on the **Power Generation** folder. The placemarked locations in this folder show power generation facilities utilizing either fossil fuel or geothermal production methods.

2. Use the Google Earth™ 3D View as well as the *Photos* layer from the *Primary Database* to examine the landscapes surrounding each placemark.

17.3 Geothermal Energy Short Answer B

B. From the perspective of landscape degradation, how do geothermal power generation sites compare with fossil fuel-based power stations?

_____
_____
_____
_____
_____
_____
_____
_____

When you complete this exploration, turn off and collapse the **17.3 Geothermal Energy** folder.

Exploration 17.4 – Tourism
Multiple Choice

Karst and hydrothermal features create relatively unique landscapes at or near the surface. Throughout human history, many of these places were utilized as spiritual, medicinal, or recreational destinations. Today, many karst or hydrothermal landscapes are protected as national parks.

Instructions for all Parts:

1. Make sure you have opened the **KMZ** file from www.mygeoscienceplace.com.

2. From the Places panel, expand **17. Karst and Hydrothermal Processes.kmz** and then open the **17.4 Tourism** folder.

Instructions for Exploration 17.4 Tourism Parts A–C:

1. Turn on and open the **Yellowstone National Park** folder. Yellowstone contains one of the world's largest concentrations of geothermal and hydrothermal activity. The five placemarks labeled *A* through *E* indicate the location of different examples of these features.'

2. Examine the landscape around each placemarked location. Turn on the *Photos* layer from the *Primary Database* to get a better view of each site. Evaluate the type of hydrothermal or geothermal feature shown at each location.

Exploration 17.4 Tourism Part A

A. Which of the placemarks in the **Yellowstone National Park** folder is the best example of a geyser?

1. A
2. B
3. C
4. D
5. E

Exploration 17.4 Tourism Part B

B. Which of the placemarks in the **Yellowstone National Park** folder is the best example of a mud pot?

1. A
2. B
3. C
4. D
5. E

Exploration 17.4 Tourism Part C

C. Which of the placemarks in the **Yellowstone National Park** folder is the best example of travertine terraces?

1. A
2. B
3. C
4. D
5. E

IMPORTANT: Be sure to turn off the *Photos* layer from the *Primary Database*.

When you complete the previous three parts, turn off and collapse the **Yellowstone National Park** folder.

Instructions for Exploration 17.4 Tourism Part D:

1. Turn on and open the **National Park Sites** (NPS) folder. Yellowstone National Park is the most well-known park for hydrothermal activity, but it is hardly the only park containing similar features. In addition, the National Park Service protects a number of sites in karst landscapes.

2. Fly to each of the placemarked parks then click the associated links to access the NPS website for each location. Use the imagery and written park descriptions to determine whether the park protects hydrothermal or karst landscapes.

Exploration 17.4 Tourism Part D

D. How many of the placemarked parks in the **National Park Sites** folder protect hydrothermal or karst landscapes?

1. 1
2. 2
3. 3
4. 4
5. 5

When you complete this exploration, turn off and collapse the **17.4 Tourism** folder.

Exploration 17.4 – Tourism
Short Answer

Instructions for all Parts:

1. Make sure you have opened the **KMZ** file from www.mygeoscienceplace.com.

2. From the Places panel, expand **17. Karst and Hydrothermal Processes.kmz** and then open the **17.4 Tourism** folder.

Instructions for Exploration 17.4 Tourism Short Answer A:

1. Open the **National Park Sites** folder.

2. The five parks shown here, along with the nearly 400 other locations that comprise the National Park System, were selected for preservation in part because they represent relatively undeveloped or undamaged examples of specific landscapes. These five sites represent karst or hydrothermal landscapes of some kind.

3. Evaluate the status of these sites, being sure to consider the balance between karst and hydrothermal sites represented.

Exploration 17.4 Tourism Short Answer A

A. The examples in the **National Park Sites** folder demonstrate that karst and hydrothermal landscapes are represented within the National Park System. Do you think one of these landscapes would be disproportionately represented over the other, considering the fact that the Park Service is less likely to protect an area that has already been commercially developed?

_____

_____

_____

_____

_____

_____

_____

Instructions for Exploration 17.4 Tourism Short Answer B:

1. Use Google Earth™ to navigate to a tourist destination that is based on karst or hydrothermal features and which was not already featured in a previous question.

2. Evaluate this location, noting the characteristics that identify it as a karst or hydrothermal landscape.

3. Use outside research, if necessary, to identify a location and to provide context for its significance as a travel destination.

Exploration 17.4 Tourism Short Answer B

B. What is the name and location, in latitude and longitude, of a significant karst or hydrothermal tourist destination? What does the site look like in Google Earth™? Provide a brief summary of the site and the features on display.

_____

_____

_____

_____

_____

_____

_____

When you complete this exploration, turn off and collapse the **17.4 Tourism** folder.

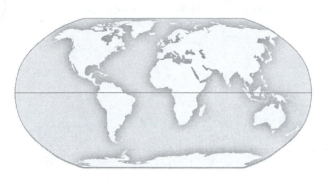

Encounter Physical Geography

Name:_____

Date: _____

Chapter 18
Topography of Arid Lands

Exploration 18.1 – Water in Waterless Regions
Multiple Choice

By definition, arid lands lack an abundance of water. Therefore, the presence of water can be particularly conspicuous. Even when water is not a permanent or frequent component of a landscape, it can play a significant role in shaping physical landforms.

Instructions for all Parts:

1. Make sure you have opened the **KMZ** file from www.mygeoscienceplace.com.

2. From the Places panel, expand **18. Topography of Arid Lands.kmz** and then open the **18.1 Water in Waterless Regions** folder.

Instructions for Exploration 18.1 Water in Waterless Regions Parts A–B:

1. Turn on and open the **Arid Streams** folder. The five placemarks labeled *A* through *E* represent a selection of water bodies.

2. Examine each of the placemarked locations, being sure to note the status of water in each landscape.

Exploration 18.1 Water in Waterless Regions Part A

A. Which of the placemarked locations in the **Arid Streams** folder would be considered an ephemeral stream?

1. A
2. B
3. C
4. D
5. E

Exploration 18.1 Water in Waterless Regions Part B

B. Which of the placemarked locations in the **Arid Streams** folder would be the best example of an exotic stream?

1. A
2. B
3. C
4. D
5. E

When you complete these parts, turn off and collapse the **Arid Streams** folder.

Instructions for Exploration 18.1 Water in Waterless Regions Part C:

1. Turn on and open the **Alluvial Fan** folder.

2. Examine each of the placemarked locations noting the presence or absence of alluvial fans at each location.

Exploration 18.1 Water in Waterless Regions Part C

C. Which of the placemarks in the **Alluvial Fan** folder is the best example of an alluvial fan?

1. A
2. B
3. C
4. D
5. E

When you complete this part, turn off and collapse the **Alluvial Fan** folder.

Instructions for Exploration 18.1 Water in Waterless Regions Part D:

1. Turn on and double-click the *Okavango* placemark to fly to a unique area in northern Botswana.

2. Examine the landform indicated by the dark patterns present at this location. Be sure to consider what processes helped to create this landform.

Exploration 18.1 Water in Waterless Regions Part D

D. Which of the following is a characteristic of many desert landscapes that best explains the presence of the feature seen in at the *Okavango* placemark?

1. absence of soil
2. impermeable surface
3. soil creep
4. interior drainage
5. mechanical weathering

When you complete this exploration, turn off and collapse the **18.1 Water in Waterless Regions** folder.

Exploration 18.1 – Water in Waterless Regions
Short Answer

Instructions for all Parts:

1. Make sure you have opened the **KMZ** file from www.mygeoscienceplace.com.

2. From the Places panel, expand **18. Topography of Arid Lands.kmz** and then open the **18.1 Water in Waterless Regions** folder.

Instructions for Exploration 18.1 Water in Waterless Regions Short Answer A:

1. Turn on and double-click the *Grand Canyon* Gigapan image to fly into a scene in Grand Canyon National Park.

2. Examine the landscape of this arid part of the American West, noting the distinctive erosional pattern present.

Exploration 18.1 Water in Waterless Regions Short Answer A

A. What is the name of the distinct erosional pattern that characterizes the landscape seen in the *Grand Canyon* Gigapan image? Describe the processes at work in this type of erosion.

_____
_____
_____
_____
_____

When you complete this part, turn off the *Grand Canyon* Gigapan image.

Instructions for Exploration 18.1 Water in Waterless Regions Short Answer B:

1. Turn on and double-click the *Water(less) Feature* placemark.

2. Examine this curious feature and the surrounding landscape in this arid region.

Exploration 18.1 Water in Waterless Regions Short Answer B

B. What type of arid land feature is shown at the *Water(less) Feature* placemark? The **Photos** layer in the **Primary Database** may be helpful for identification purposes. Provide an explanation for the existence of this waterless feature. Be sure to indicate how it was formed as well as why it appears the way it does.

_____

_____

_____

_____

_____

When you complete this exploration, turn off and collapse the **18.2 Water in Waterless Regions** folder.

Exploration 18.2 – Work of the Wind
Multiple Choice

Although wind does not impact arid landscapes on the magnitude water does, it is responsible for transient features such as sand dunes. The morphology of dunes is largely a reflection of wind direction combined with the volume of available sand or sediment.

Instructions for all Parts:

1. Make sure you have opened the **KMZ** file from www.mygeoscienceplace.com.

2. From the Places panel, expand **18. Topography of Arid Lands.kmz** and then open the **18.2 Work of the Wind** folder.

Instructions for Exploration 18.2 Work of the Wind Parts A–B:

1. Turn on and open the **Dune Types** folder. The five placemarks labeled *A* through *E* provide examples of different dune types.

2. Examine each of the placemarks, being sure to identify the type of dune present at each location.

Exploration 18.2 Work of the Wind Part A

A. Which of the placemarks in the **Dune Types** folder best represents a barchan dune?

1. A
2. B
3. C
4. D
5. E

Exploration 18.2 Work of the Wind Part B

B. Which of the placemarks in the **Dune Types** folder best represents a star dune?

1. A
2. B
3. C
4. D
5. E

When you complete the previous two parts, turn off and collapse the **Dune Types** folder.

Instructions for Exploration 18.2 Work of the Wind Part C:

1. Turn on and double-click the *Namib Desert* Gigapan image to fly into a landscape near the Atlantic coast of Namibia.

2. Evaluate the landscape shown in the image noting the type of desert represented at this location.

Exploration 18.2 Work of the Wind Part C

C. The landscape seen in the *Namib Desert* image is best described by which of the following desert types?

1. erg
2. reg
3. Hamada
4. pediment
5. desert pavement

When you complete this part, turn off the *Namib Desert* Gigapan image.

Instructions for Exploration 18.2 Work of the Wind Part D:

1. Turn on and double-click the *Prevailing Wind Direction* placemark to fly into a desert landscape in Qatar.

2. Evaluate the dunes seen at this location, paying particular attention to the type of dunes present and the likely prevailing wind direction. Be sure to remember that winds are classified by the direction from which they come.

Exploration 18.2 Work of the Wind Part D

D. Which of the following prevailing wind directions is most likely responsible for the dunes seen at the *Prevailing Wind Direction* placemark?

1. southeast
2. northwest
3. northeast
4. southeast
5. south

When you complete this exploration, turn off and collapse the **18.2 Work of the Wind** folder.

Exploration 18.2 – Work of the Wind
Short Answer

Instructions for all Parts:

1. Make sure you have opened the **KMZ** file from www.mygeoscienceplace.com.

2. From the Places panel, expand **18. Topography of Arid Lands.kmz** and then open the **18.2 Work of the Wind** folder.

Instructions for Exploration 18.2 Work of the Wind Short Answer A:

1. Double-click the NASA icon on the **Dust Storm in the Simpson Desert, Australia** folder, then select the *Overlay On* option.

2. The overlay shows a MODIS satellite image of a dust storm that occurred in the Simpson Desert of central Australia in September 2006. Click the folder hyperlink to learn more about this extreme weather event.

3. Turn on and double-click the *Simpson Dunes* placemark to zoom in to a portion of the Simpson Desert.

4. Examine the patterns in the dunes seen in the Google Earth™ imagery and compare those patterns with the patterns seen in the imagery of the dust storm.

Exploration 18.2 Work of the Wind Short Answer A

A. Based on the evidence in the image overlay, in what direction was the wind blowing during the dust storm in September 2006? What types of dunes are present at the placemark in the Simpson Desert? What is the relationship between the types of dunes seen here and the wind direction evident in the dust storm?

_____

_____

_____

_____

_____

When you complete this part, turn off and collapse the **Dust Storm in the Simpson Desert, Australia** folder.

Instructions for Exploration 18.2 Work of the Wind Short Answer B:

1. Turn on and double-click the *Namib Desert* Gigapan image to fly to this desert area along the Atlantic coast of southern Africa.

2. Examine the local landscape, being sure to note the characteristics of this desert and the possible weather patterns that might contribute to its formation.

Exploration 18.2 Work of the Wind Short Answer B

B. Explain the characteristics of local weather patterns that would result in the dune formations seen in the *Namib Desert* image. Be sure to indicate the type of dunes present and any relevant details of wind intensity or direction.

_____

_____

_____

_____

_____

When you complete this exploration, turn off and collapse the **18.2 Work of the Wind** folder.

Exploration 18.3 – Arid Landforms
Multiple Choice

The world's arid lands are home to distinct landforms and landform assemblages that reflect the overriding environmental factor of dryness. Basin-and-range and mesa-and-scarp terrain are two of the more common landform assemblages that are repeated around the world under dry conditions.

Instructions for all Parts:

1. Make sure you have opened the **KMZ** file from www.mygeoscienceplace.com.

2. From the Places panel, expand **18. Topography of Arid Lands.kmz** and then open the **18.3 Arid Landforms** folder.

Instructions for Exploration 18.3 Arid Landforms Parts A–B:

1. Turn on and open the **Unique Arid Landforms** folder. The five placemarks in this folder show examples of landforms often found in arid locations.

2. Examine the landforms found at each placemark, noting the specific type of feature present at each location. For placemark *D*, fly to the placemark then turn on the *Photos* layer from the *Primary Database* to examine the specific feature found at this location.

Exploration 18.3 Arid Landforms Part A

A. Which of the following arid land features is best represented by placemark *C* in the **Unique Arid Landforms** folder?

1. mesa
2. arch
3. plateau
4. inselberg
5. bornhardt

Exploration 18.3 Arid Landforms Part B

B. Which of the following arid land features is best represented by placemark *E* in the **Unique Arid Landforms** folder?

1. mesa
2. arch
3. plateau
4. inselberg
5. bornhardt

IMPORTANT: Be sure to turn off the *Photos* layer from the *Primary Database*.

When you complete these parts, turn off and collapse the **Unique Arid Landforms** folder.

Instructions for Exploration 18.3 Arid Landforms Part C:

1. Turn on and open the **Basin-and-Range** folder.

2. Examine each of the placemarked landscapes, being sure to note the presence or absence of a basin-and-range structure at each location.

Exploration 18.3 Arid Landforms Part C

C. Which of the following placemarks in the **Basin-and-Range** folder best represents basin-and-range topography?

1. A
2. B
3. C
4. D
5. E

When you complete this part, turn off and collapse the **Basin-and-Range** folder.

Instructions for Exploration 18.3 Arid Landforms Part D:

1. Turn on and open the **Badlands** folder.

2. Examine each of the placemarked landscapes, being sure to note the presence or absence of badlands at each location.

Exploration 18.3 Arid Landforms Part D

D. Which of the following placemarks in the **Badlands** folder does **not** show badlands topography?

1. A
2. B
3. C
4. D
5. E

When you complete this exploration, turn off and collapse the **18.3 Arid Landforms** folder.

Exploration 18.3 – Arid Landforms
Short Answer

Instructions for all Parts:

1. Make sure you have opened the **KMZ** file from www.mygeoscienceplace.com.

2. From the Places panel, expand **18. Topography of Arid Lands.kmz** and then open the **18.3 Arid Landforms** folder.

Instructions for Exploration 18.3 Arid Landforms Short Answer A:

1. Turn on and open the **Compare and Contrast** folder, then double-click placemark *1* to fly into a desert landscape.

2. Evaluate the arid landscape found at this location.

3. Double-click placemark *2* and repeat the same evaluation, being sure to note the differences in desert characteristics between the two placemarks.

Exploration 18.3 Arid Landforms Short Answer A

A. Describe the different desert landscapes found in the placemarks in the *Compare and Contrast* folder. Explain how these patterns may have developed, making sure to use terminology appropriate for deserts and arid lands.

_____

_____

_____

_____

_____

When you complete this part, turn off and collapse the **Compare and Contrast** folder.

Instructions for Exploration 18.3 Arid Landforms Short Answer B:

1. Turn on and double-click the *Death Valley* placemark to fly into this national park on the California and Nevada border.

2. Examine the landscape, being sure to note any characteristics or features common to deserts and arid lands.

Exploration 18.3 Arid Landforms Short Answer B

B. Provide a brief description of the desert and arid land features present in the landscape found at the *Death Valley* placemark.

_____

_____

_____

_____

_____

When you complete this exploration, turn off and collapse the **18.3 Arid Landforms** folder.

Exploration 18.4 – Humans in Arid Lands
Multiple Choice

Despite the challenges presented by arid lands, humans have inhabited the world's dry lands since the establishment of the earliest civilizations. Living in these places requires special adaptations and technologies. Though modification of arid lands is possible, and in some areas increasing in frequency, these sensitive environments may not be well suited to large-scale population pressures.

Instructions for all Parts:

1. Make sure you have opened the **KMZ** file from www.mygeoscienceplace.com.

2. From the Places panel, expand **18. Topography of Arid Lands.kmz** and then open the **18.4 Humans in Arid Lands** folder.

Instructions for Exploration 18.4 Humans in Arid Lands Part A:

1. Turn on and double-click the *Landscape Change* placemark to fly into an arid landscape in Tunisia.

2. The Historical Imagery time slider will set the view to one showing this location in July 2009. Use the time slider to change the imagery date to January 1987.

3. Compare the imagery at the *Landscape Change* placemark from 2009 and 1987, being sure to note any significant changes that have occurred at this location. Zoom in as necessary to examine details of the landscape change.

Exploration 18.4 Humans in Arid Lands Part A

A. Which of the following statements is best supported by the evidence seen at the *Landscape Change* placemark?

1. The area experienced a severe forest fire between 1987 and 2009.
2. The area has been severely eroded by flooding.
3. The area has been fenced and vegetation has started to recover.
4. The area has been fenced and has been subjected to intense grazing.
5. The area has been overrun by migrating sand dunes.

IMPORTANT: Be sure to turn off the Historical Imagery slider.

When you complete this part, turn off the *Landscape Change* placemark.

Instructions for Exploration 18.4 Humans in Arid Lands Part B:

1. Double-click the **Las Vegas** folder, then turn on the *1973* image overlay.

2. Evaluate the spatial footprint of Las Vegas and its suburbs.

3. Turn on the *2006* image overlay and repeat the same evaluation of the spatial extent of the Las Vegas area.

Exploration 18.4 Humans in Arid Lands Part B

B. Based on the evidence in the **Las Vegas** folder, what is the best characterization for the change in size of the Las Vegas metropolitan area between 1973 and 2006?

1. substantial decrease
2. moderate decrease
3. no change
4. moderate increase
5. substantial increase

When you complete this part, turn off and collapse the Las Vegas folder.

Instructions for Exploration 18.4 Humans in Arid Lands Part C:

1. Turn on and double-click the *Irrigation* placemark to zoom to a location in the Sahara Desert near the border of Egypt and Sudan.

2. Evaluate the features found at this location.

Exploration 18.4 Humans in Arid Lands Part C

C. Which of the following statements is best supported by the evidence seen at the *Irrigation* placemark?

1. Egypt has created residential pods in the eastern Sahara Desert.
2. This image demonstrates that food can be grown in any arid setting.
3. Food is being produced at this location using surface irrigation from the Nile River.
4. Humans have established areas of agricultural production in the heart of the eastern Sahara Desert.
5. No evidence of human settlement shows in this location.

When you complete this part, turn off the *Irrigation* placemark.

Instructions for Exploration 18.4 Humans in Arid Lands Part D:

1. Double-click the **Lake Faguibine, Mali** folder to fly to an area of central Mali that has experienced desertification over the past few decades.

2. Turn on the *1978* image overlay and evaluate the extent of the lake at this time, paying particular attention to the size of the east-west dimension.

3. Repeat the same evaluation of the lake's extent using the *1987, 2001,* and *2006* image layers.

Exploration 18.4 Humans in Arid Lands Part D

D. Based on the evidence in the **Lake Faguibine, Mali** folder, what has been the change in the east-west extent of Lake Faguibine from 1978 to 2006?

1. An increase of 25 km.
2. No change in dimension.
3. A decrease of 5 km.
4. A decrease of 25 km.
5. A decrease of 50 km.

IMPORTANT: Be sure to turn off the Historical Imagery.

When you complete this exploration, turn off and collapse the **18.4 Humans in Arid Lands** folder.

Exploration 18.4 – Humans in Arid Lands
Short Answer

Instructions for all Parts:

1. Make sure you have opened the **KMZ** file from www.mygeoscienceplace.com.

2. From the Places panel, expand **18. Topography of Arid Lands.kmz** and then open the **18.4 Humans in Arid Lands** folder.

Instructions for Exploration 18.4 Humans in Arid Lands Short Answer A:

1. Double-click the **Lake Hamoun, Iran** folder to fly to an area of eastern Iran near the border of Afghanistan.

2. Examine the landscape of this region using the default Google Earth™ imagery. Pay particular attention to the density and location of settlements.

3. Turn on the *1976* image overlay to see the extent of Lake Hamoun at this time. Be sure to note the patterns of development seen around the lake.

4. Repeat the same evaluation using the *2001* image layer, noting any significant changes between the two Landsat images and the default Google Earth™ imagery.

Exploration 18.4 Humans in Arid Lands Short Answer A

A. Based on the evidence in the **Lake Hamoun, Iran** folder, what has happened to the lake since the mid-1970s? Describe the changes in the physical environment and indicate how those changes may have impacted life in the communities of this region.

_____
_____
_____
_____
_____

When you complete this part, turn off and collapse the **Lake Hamoun, Iran** folder.

Instructions for Exploration 18.4 Humans in Arid Lands Short Answer B:

1. Double-click the **Las Vegas** folder and then turn on the *1973* image overlay.

2. Evaluate the spatial footprint of Las Vegas and its suburbs.

3. Turn on the *2006* image overly and repeat the same evaluation of the spatial extent of the Las Vegas area. Be sure to consider the impact the growth may have in this arid part of the Intermountain West. Utilize the default Google Earth™ imagery to see the current extent of the area.

18.4 Humans in Arid Lands Short Answer B

B. Do you feel the growth indicated by the imagery in the **Las Vegas** folder is a sustainable pattern of development for urban locations in such dry environments? What are methods or actions that might be utilized to help conserve the supply of available water?

_____

_____

_____

_____

_____

_____

_____

When you complete this exploration, turn off and collapse the **18.4 Humans in Arid Lands** folder.

Name:_____

Date: _____

Chapter 19
Glacial Modification of Terrain

Exploration 19.1 – Glaciers Past & Present
Multiple Choice

Glaciers and their modification of terrain have waxed and waned over the Earth's geologic past. Glacial ice has a remarkable ability to sculpt landscapes into distinctive landforms through erosional abilities. A second group of landforms is created by the depositional capabilities of glaciers.

Instructions for all Parts:

1. Make sure you have opened the **KMZ** file from the www.mygeoscienceplace.com.

2. From the Places panel, expand **19. Glacial Processes.kmz** and then open the **19.1 Glaciers Past and Present** folder.

Instructions for Exploration 19.1 Glaciers Past and Present Parts A–B:

1. Turn on and open the **Glacier Types** folder.

2. View placemarks *A* through *E* with the different types of glaciers in mind.

Exploration 19.1 Glaciers Past and Present Part A

A. Which of the placemarked locations in the **Glacier Types** folder shows a cirque glacier?

1. A
2. B
3. C
4. D
5. E

Exploration 19.1 Glaciers Past and Present Part B

B. Which of the placemarked locations in the **Glacier Types** folder shows a piedmont glacier?

1. A
2. B
3. C
4. D
5. E

When you complete these parts, turn off and collapse the **Glacier Types** folder.

Instructions for Exploration 19.1 Glaciers Past and Present Part C:

1. Turn on and double-click the *Past Glacial Processes* placemark.

2. Examine the landscape around the placemark, being sure to consider the visible impacts of past glacial and periglacial processes.

Exploration 19.1 Glaciers Past and Present Part C

C. Identify the statement that is most strongly supported by the evidence seen in the landscape associated with the *Past Glacial Processes* placemark.

1. This location shows the impact of sea-level change during the Pleistocene Glaciation.
2. This location is a remnant of a Pleistocene Era pluvial lake, which has subsequently been reduced in size.
3. This location shows the impact of scouring of a continental ice sheet.
4. This location is an area where a highland ice cap has recently melted due to climate change.
5. This location shows the impact of alpine glacial erosion.

When you complete this part, turn off the *Past Glacial Processes* placemark.

Instructions for Exploration 19.1 Glaciers Past and Present Part D:

1. Turn on and open the **Pleistocene Glaciation** folder.

2. Examine placemarks *A* through *E* with special consideration given to their general continental locations.

Exploration 19.1 Glaciers Past and Present Part D

D. Which of the placemarked locations in the folder would **not** have been covered by a continental ice sheet during Pleistocene Glaciation?

1. A
2. B
3. C
4. D
5. E

When you complete this exploration, turn off and collapse the **19.1 Glaciers Past and Present** folder.

Exploration 19.1 – Glaciers Past and Present
Short Answer

Instructions for all Parts:

1. Make sure you have opened the **KMZ** file from the www.mygeoscienceplace.com.

2. From the Places panel, expand **19. Glacial Processes.kmz** and then open the **19.1 Glaciers Past and Present** folder.

Instructions for Exploration 19.1 Glaciers Past and Present Short Answer A:

1. Turn on and double-click the *Missoula Floods* placemark.

2. View the landscape shown at the *Missoula Floods* placemark related to a series of events known as the Missoula Floods.

3. Utilize outside sources and research these events.

Exploration 19.1 Glaciers Past and Present Short Answer A

A. Provide a summary of what happened during the Missoula Floods and how the unique landscape at the *Missoula Floods* placemark is related to those events.

_____

_____

_____

_____

_____

When you complete this part, turn off the *Missoula Floods* placemark.

Instructions for Exploration 19.1 Glaciers Past and Present Short Answer B:

1. Turn on and double-click the **Larsen B Ice Shelf** folder to view satellite imagery obtained shortly after the 2002 collapse of a portion of Antarctica's Larsen Ice Shelf.

2. Examine the local impacts of this event. You may want to compare this image to contemporary Google Earth™ imagery as a part of your examination.

3. Utilize outside sources to research the broader causes and effects of the Larsen Ice Shelf collapse.

Exploration 19.1 Glaciers Past and Present Short Answer B

B. What were the rough dimensions of the Larsen B Ice Shelf that collapsed? How does this phenomenon relate to global climate? Is this an isolated incident or do other examples exist? Use the images provided and any outside sources to formulate your response.

_____

_____

_____

_____

_____

When you complete this exploration, turn off and collapse the **19.1 Glaciers Past and Present** folder.

Exploration 19.2 – Glacier Formation, Movement, and Effects
Multiple Choice

Generally, we expect glaciers to accumulate and advance during cooler and wetter climatic periods of the Earth's history. Conversely, in warmer periods glaciers typically melt and retreat. Both glacial advance and retreat correlate with distinctive processes and landforms.

Instructions for all Parts:

1. Make sure you have opened the **KMZ** file from the www.mygeoscienceplace.com.

2. From the Places panel, expand **19. Glacial Processes.kmz** and then open the **19.2 Glacier Formation, Movement, and Effects** folder.

Instructions for Exploration 19.2 Glacier Formation, Movement, and Effects Part A:

1. Turn on and double-click the *McCarty Glacier* placemark.

2. Click the placemark hyperlink to view images maintained by the National Snow and Ice Data Center. Close the pop-up window when you are done examining the historical images.

3. Utilize the time slider to assess the advance or retreat of the McCarty Glacier between 9/01/1996 and 8/07/2005.

Exploration 19.2 Glacier Formation, Movement, and Effects Part A

A. Based on the historical imagery at the location in the folder, how far did the glacier advance or retreat between 1996 and 2005?

1. advanced approximately 100 meters
2. advanced approximately 750 meters
3. retreated approximately 750 meters
4. retreated approximately 1.5 kilometers
5. retreated approximately 5 kilometers

IMPORTANT: Be sure to turn off the Historical Imagery time slider.

When you have completed this part, turn off the *McCarty Glacier* placemark.

Instructions for Exploration 19.2 Glacier Formation, Movement, and Effects Part B:

1. Turn on and double-click the **Equilibrium Line** folder.

2. Examine the glacier seen in this image along with the colored lines that represent hypothetical lines of equilibrium.

Exploration 19.2 Glacier Formation, Movement, and Effects Part B

B. Which of the colored lines shown in the **Equilibrium Line** folder is the best choice for the location of the equilibrium line for this glacier?

1. A
2. B
3. C
4. D
5. E

When you have completed this part, turn off and collapse the **Equilibrium Line** folder.

Instructions for Exploration 19.2 Glacier Formation, Movement, and Effects Part C:

1. Turn on and double-click the **Helheim Glacier** folder and examine the images taken on different dates, considering glacial flow versus glacial advance or retreat.

2. Examine the imagery for 1986, 2003, and 2006 in light of your knowledge of glacial movement.

Exploration 19.2 Glacier Formation, Movement, and Effects Part C

C. Based on your evaluation of imagery in the **Helheim Glacier** folder, identify the statement that is most strongly supported.

1. The flow of this glacier is dictated by the westerly winds.
2. This glacier is flowing in a predominantly west to east direction but it is retreating in an east to west direction.
3. This glacier is retreating in a general west to east fashion.
4. The glacier is flowing from a higher point to a lower point but is advancing from a lower point to a higher point.
5. This glacier is flowing east to west due to rising sea levels.

When you have completed this part, turn off and collapse the **Helheim Glacier** folder.

Instructions for Exploration 19.2 Glacier Formation, Movement, and Effects Part D:

1. Turn on and double-click the **Terminology** folder and examine this glacial landscape.

2. Use the navigation capabilities of Google Earth™ to examine each placemarked location on the landscape.

Exploration 19.2 Glacier Formation, Movement, and Effects Part D

D. Which of the following terms associated with glacial processes is **<u>not</u>** represented by location(s) placemarked in the **Terminology** folder?

1. abrasion
2. melt stream
3. glacio-fluvio deposition
4. drift
5. basal slip

When you complete this exploration, turn off and collapse the **19.2 Glacier Formation, Movement, and Effects** folder.

Exploration 19.2 – Glacier Formation, Movement, and Effects
Short Answer

Instructions for all Parts:

1. Make sure you have opened the **KMZ** file from the www.mygeoscienceplace.com.

2. From the Places panel, expand **19. Glacial Processes.kmz** and then open the **19.2 Glacier Formation, Movement, and Effects** folder.

Instructions for Exploration 19.2 Glacier Formation, Movement, and Effects
Short Answer A:

1. Turn on and double-click the *McCarty Glacier* placemark, then click the placemark hyperlink to examine repeat photography images of this glacial landscape.

2. Compare these images showing the changes since the early 20th century to changes of the same glaciers you can see using the historical imagery in the Google Earth™ 3D Viewer.

Exploration 19.2 Glacier Formation, Movement, and Effects Short Answer A

A. Provide a summary of the changes that have occurred at the McCarty Glacier over the past few decades in comparison to the changes that have been seen over the past 100 years. Can you determine whether the rate of change has increased or decreased in recent years? What advantages and disadvantages do aerial images have over ground-based images?

_____

_____

_____

_____

_____

When you have completed this part, turn off the *McCarty Glacier* placemark and turn off the Historical Imagery.

Instructions for Exploration 19.2 Glacier Formation, Movement, and Effects Short Answer B:

1. Turn on and double-click the **Kangerdlugssuaq Glacier** folder and examine the 1991 and 2007 image sets along with the dataset prepared by Stearns. The lines on the Stearns image show the location of the toe of the glacier at various years. Be sure to note the change in glacial extent between 2005 and 2007 and compare that trend to the most current default Google Earth™ imagery.

Exploration 19.2 Glacier Formation, Movement, and Effects Short Answer B

B. What factors might be contributing to the changes in the flow of the glacier, particularly its apparent advancement from 2005 to 2007 during a time of widespread global glacial retreat.

_____

_____

_____

_____

_____

When you complete this exploration, turn off and collapse the **19.2 Glacier Formation, Movement, and Effects** folder.

Exploration 19.3 – Continental Ice Sheets
Multiple Choice

Continental ice sheets are unmatched in their scale and capacity to alter large regions of the planet's surface. The last great advance of continental ice occurred during the Pleistocene when much of North America was covered in several thousand feet of ice. These ice sheets leveled much of the landscape and were responsible for excavating a number of prominent features.
Instructions for all Parts:

1. Make sure you have opened the **KMZ** file from the www.mygeoscienceplace.com.

2. From the Places panel, expand **19. Glacial Processes.kmz** and then open the **19.3 Continental Ice Sheets** folder.

Instructions for Exploration 19.3 Continental Ice Sheets Part A:

1. Turn on and double-click the *Ehalkivi* placemark.

2. Examine the Google Earth™ imagery, noting the relative latitudinal position of this location.

3. Open the *Photos* layer from the *Primary Database* and examine some nearby images to get a better view of the indicated feature and the surrounding landscape.

Exploration 19.3 Continental Ice Sheets Part A

A. What glacial term best describes the landscape feature highlighted at the *Ehalkivi* placemark?

1. roche moutonée
2. glacial flour
3. terminal moraine
4. glacial erratic
5. kettle

IMPORTANT: Be sure to turn off the *Photos* layer from the *Primary Database*.

When you have completed this part, turn off the *Ehalkivi* placemark.

Instructions for Exploration 19.3 Continental Ice Sheets Part B:

1. Turn on and double-click the *Features 1* placemark and attempt to determine the glacial landforms present on the landscape.

2. Turn on and double-click the *Features 2* placemark for a different perspective.

Exploration 19.3 Continental Ice Sheets Part B

B. Identify the glacial landforms illustrated by the *Features 1* and *Features 2* placemarks.

1. kettles
2. drumlins
3. glacial erratics
4. deltas
5. ice blocks

IMPORTANT: Turn off the Historical Imagery when you complete this part.

When you have completed this part, turn off the *Features 1* and *Features 2* placemarks.

Instructions for Exploration 19.3 Continental Ice Sheets Part C:

1. Turn on and double-click the *Continental Ice Sheet Feature* path to fly into part of southern Wisconsin.
2. Examine the glacial landform that roughly parallels the white line to the east and southeast.

Exploration 19.3 Continental Ice Sheets Part C

C. What glacial landform parallels the eastern side of the *Continental Ice Sheet Feature* path?

1. medial moraine
2. horn
3. arête
4. drumlin
5. terminal moraine

When you have completed this part, turn off the *Continental Ice Sheet Feature* path.

Instructions for Exploration 19.3 Continental Ice Sheets Part D:

1. Turn on and double-click the *Roche Moutonnée* placemark.

2. Determine the direction of ice movement at this location.

Exploration 19.3 Continental Ice Sheets Part D

D. What is the direction of ice movement at the *Roche Moutonnée* placemark?

1. east to west
2. west to east
3. south to north
4. north to south
5. lower to higher elevation

When you complete this exploration, turn off and collapse the **19.3 Continental Ice Sheets** folder.

Exploration 19.3 – Continental Ice Sheets
Short Answer

Instructions for all Parts:

1. Make sure you have opened the **KMZ** file from the www.mygeoscienceplace.com

2. From the Places panel, expand **19. Glacial Processes.kmz** and then open the **19.3 Continental Ice Sheets** folder.

Instructions for Exploration 19.3 Continental Ice Sheets Short Answer A:

1. Turn on and double-click the *Finger Lakes* placemark to fly to this region in upstate New York.

2. Examine the available imagery to evaluate the regional landscape around the Finger Lakes. Note the relative latitudinal position of this location and the possible impacts glaciers had on the creation of this landscape.

3. Use your textbook or outside sources to get a better understanding of this landscape.

Exploration 19.3 Continental Ice Sheets Short Answer A

A. Utilizing your Google Earth™ assessment of the Finger Lakes along with any outside research, provide a brief explanation of the creation of the Finger Lakes. Be sure to address the lakes' orientation in your response.

_____
_____
_____
_____
_____

When you have completed this part, turn off the *Finger Lakes* placemark.

Instructions for Exploration 19.3 Continental Ice Sheets Short Answer B:

1. Turn on and double-click the *Greenland Pools* placemark.

2. Examine the imagery, surrounding landscape, and relative latitudinal position of this location.

Exploration 19.3 Continental Ice Sheets Short Answer B

B. What is significant about the blue features illustrated in the *Greenland Pools* placemark? What does the presence of these pools suggest? Complete some outside research to briefly elaborate on the process that is illustrated.

_____
_____
_____
_____
_____

When you complete this exploration, turn off and collapse the **19.3 Continental Ice Sheets** folder.

Exploration 19.4 – Mountain Glaciers
Multiple Choice

Far smaller than their continental counterparts, mountain glaciers alter the landscape of highland areas as they accumulate in high-altitude basins and are channeled through mountain valleys. Beyond transforming V-shaped stream valleys into U-shaped glacial valleys, alpine glaciers are often responsible for creating some of the world's most spectacular mountain landscapes.

Instructions for all Parts:

1. Make sure you have opened the **KMZ** file from the www.mygeoscienceplace.com.

2. From the Places panel, expand **19. Glacial Processes.kmz** and then open the **19.4 Mountain Glaciers** folder.

Instructions for Exploration 19.4 Mountain Glaciers Parts A–B:

1. Turn on and open the **Alpine Landforms** folder.

2. Double-click placemarks *A* through *E*, evaluating the alpine glacial landforms marked by each placemark. Use the navigational capabilities of Google Earth™ to view each landform from a variety of perspectives.

Exploration 19.4 Mountain Glaciers Part A

A. Which of the placemarks in the **Alpine Landforms** folder represents an arête?

1. A
2. B
3. C
4. D
5. E

Exploration 19.4 Mountain Glaciers Part B

B. Which of the placemarks in the **Alpine Landforms** folder represents a tarn?

1. A
2. B
3. C
4. D
5. E

When you have completed these parts, turn off and collapse the **Alpine Landforms** folder.

Instructions for Exploration 19.4 Mountain Glaciers Part C:

1. Turn on and double-click the *Glacial Deposition* placemark.

2. Examine the forested curvilinear feature that is placemarked in this glacial landscape.

Exploration 19.4 Mountain Glaciers Part C

C. Which of the following terms best identifies the feature located at the *Glacial Deposition* placemark?

1. lateral moraine
2. medial moraine
3. recessional moraine
4. roche moutonnée
5. cirque

When you have completed this part, turn off the *Glacial Deposition* placemark.

Instructions for Exploration 19.4 Mountain Glaciers Part D:

1. Turn on and double-click the **Lateral Moraine** folder.

2. Double-click placemarks *A* through *E*, evaluating the placemarked locations as potential lateral moraines.

Exploration 19.4 Mountain Glaciers Part D

D. Which of the placemarks in the **Lateral Moraine** folder represents a lateral moraine?

1. A
2. B
3. C
4. D
5. E

When you have completed this exploration, turn off and collapse the **19.4 Mountain Glaciers** folder.

Exploration 19.4 – Mountain Glaciers
Short Answer

Instructions for all Parts:

1. Make sure you have opened the **KMZ** file from the www.mygeoscienceplace.com.

2. From the Places panel, expand **19. Glacial Processes.kmz** and then open the **19.4 Mountain Glaciers** folder.

Instructions for Exploration 19.4 Mountain Glaciers Short Answer A:

1. Turn on and double-click the *Importance of Glaciers* placemark.

2. Use the navigational capabilities of Google Earth™ to evaluate this landscape, noting the physical and human components and their potential relationship.

Exploration 19.4 Mountain Glaciers Short Answer A

A. Describe the physical and human components of the landscape near the *Importance of Glaciers* placemark. What role do you think the glacier plays for people of this region?

_____

_____

_____

_____

_____

Instructions for Exploration 19.4 Mountain Glaciers Short Answer B:

1. Turn on and double-click the *Grand Teton National Park* Gigapan image.

2. Zoom in to the image, noting alpine glacial landforms.

Exploration 19.4 Mountain Glaciers Short Answer B

B. Describe at least three alpine glacial landforms that can be seen in the *Grand Teton National Park* Gigapan image.

_____

_____

_____

_____

_____

When you have completed this exploration, turn off and collapse the **19.4 Mountain Glaciers** folder.

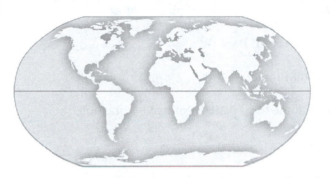

Encounter Physical Geography

Name:_____

Date: _____

Chapter 20
Coastal Processes and Terrain

Exploration 20.1 – Waves and Tidal Action
Multiple Choice

The world's coastlines are locations of constant erosion and deposition. Landforms such as rugged headlands and sandy beaches are testament to these ongoing processes.

Instructions for all Parts:

1. Make sure you have opened the **KMZ** file from www.mygeoscienceplace.com.

2. From the Places panel, expand **20. Coastal Processes and Terrain.kmz** and then open the **20.1 Waves and Tidal Action** folder.

Instructions for Exploration 20.1 Waves and Tidal Action Part A:

1. Turn on and double-click the *Longshore A* placemark. Examine the evidence of longshore currents for the selected date at this location.

2. Turn on and double-click the *Longshore B* placemark. Examine the evidence of longshore currents for the selected date at this location.

Exploration 20.1 Waves and Tidal Action Part A

A. Which of the following statements regarding longshore current activity at the *Longshore A* placemark and the *Longshore B* placemark is most strongly supported?

1. Longshore currents at the *Longshore A* placemark will transport material southward down the beach, while longshore currents at the *Longshore B* placemark will transport material northward up the beach.
2. Longshore currents at the *Longshore A* placemark will transport material northward up the beach, while longshore currents at the *Longshore B* placemark will transport material southward down the beach.
3. Longshore currents at both the *Longshore A* and *Longshore B* placemarks will transport material northward up the beach.
4. Longshore currents at both the *Longshore A* and *Longshore B* placemarks will transport material southward down the beach.
5. Longshore currents cannot be evaluated visually; wind data is necessary.

IMPORTANT: Be sure to turn off the Historical Imagery.

When you complete this part, turn off the *Longshore A* and *Longshore B* placemarks.

Instructions for Exploration 20.1 Waves and Tidal Action Part B:

1. Turn on and double-click the **Wave Refraction** folder.

2. Evaluate this coastal landscape, taking note of the local patterns of wave refraction.

Exploration 20.1 Waves and Tidal Action Part B

B. Which placemarked location in the **Wave Refraction** folder is subject to the greatest degree of wave-based erosion?

1. A
2. B
3. C
4. D
5. E

When you complete this part, turn off and collapse the **Wave Refraction** folder.

Instructions for Exploration 20.1 Waves and Tidal Action Part C:

1. Turn on and double-click the **Tidal Variation** folder.

2. Evaluate the landscape, taking note of the potential for tidal variations in water levels.

Exploration 20.1 Waves and Tidal Action Part C

C. Which placemarked location in the **Tidal Variation** folder would likely exhibit the greatest increase in exposed sediment area at low tide?

1. A
2. B
3. C
4. D
5. E

When you complete this part, turn off and collapse the **Tidal Variation** folder.

Instructions for Exploration 20.1 Waves and Tidal Action Part D:

1. Turn on and double-click the *2011 Tsunami Amplitude* layer.

2. Evaluate the patterns of wave height generated by the 2011 tsunami that hit Japan.

Exploration 20.1 Waves and Tidal Action Part D

D. Which statement is most strongly supported by the data seen in the *2011 Tsunami Amplitude* layer?

1. Kyushu experienced the highest wave heights of the Japanese islands.
2. Wave heights were much higher in the Sea of Japan than western North Pacific Ocean.
3. No US locations experienced wave heights in excess of 40 cm.
4. The area with a wave height exceeding 200 cm extended more than 400 km from the Japanese coast.
5. The imagery suggests that Indonesia was the most severely affected country during the tsunami.

When you have completed this exploration, turn off and collapse the **20.1 Waves and Tidal Action** folder.

Exploration 20.1 – Waves and Tidal Action
Short Answer

Instructions for all Parts:

1. Make sure you have opened the **KMZ** file from www.mygeoscienceplace.com.

2. From the Places panel, expand **20. Coastal Processes and Terrain.kmz** and then open the **20.1 Waves and Tidal Action** folder.

Instructions for Exploration 20.1 Waves and Tidal Action Short Answer A:

1. Turn on and double-click the *California Coast* placemark. Examine the landscape of this coastal location, noting any impacts of wave refraction.

Exploration 20.1 Waves and Tidal Action Short Answer A

A. Is the *California Coast* placemark pinned to a location of relatively higher or lower wave energy? What about the coastline area immediately to its north that has a broad sandy beach? Eventually, what will happen to the headland area near the placemark?

_____
_____
_____
_____
_____

Instructions for Exploration 20.1 Waves and Tidal Action Short Answer B:

1. Double-click the *Brant Beach* placemark to fly to this Jersey Shore town, located on Long Beach Island.

2. Examine the scene, noting any evidence of beach drift as well as any human modification of that process.

Exploration 20.1 Waves and Tidal Action Short Answer B

B. Based on the angle of the shoreline, the direction of the waves, and the accumulation of beach sand seen here, in what direction is beach drift likely occurring? Be sure to indicate how human activity is affecting that process.

_____

_____

_____

_____

_____

When you have completed this exploration, turn off and collapse the **20.1 Waves and Tidal Action** folder.

Exploration 20.2 – Coastal Landforms
Multiple Choice

One beach is often quite different than the next. Factors such as wave energy and currents, the physical makeup of headlands, and the slope of coasts contribute to differences in coastal deposition and the resulting landforms.

Instructions for all Parts:

1. Make sure you have opened the **KMZ** file from www.mygeoscienceplace.com.

2. From the Places panel, expand **20. Coastal Processes and Terrain**.kmz and then open the **20.2 Coastal Landforms** folder.

Instructions for Exploration 20.2 Coastal Landforms Part A:

1. Turn on and double-click the **Drowned River Valleys** folder to view an estuary called Chesapeake Bay.

2. To get a better understanding of the submerged river valleys and their original channels within this estuary, turn on the **Chesapeake Bay** folder. This imagery illustrates the bathymetry, or underwater topography, of Chesapeake Bay.

Exploration 20.2 Coastal Landforms Part A

A. Which of the placemarked locations in the **Drowned River Valleys** folder is **not** located in a clearly defined channel?

1. A
2. B
3. C
4. D
5. E

When you complete this part, turn off and collapse the **Drowned River Valleys** folder and turn off the **Chesapeake Bay** folder.

Instructions for Exploration 20.2 Coastal Landforms Parts B–C:

1. Turn on and expand the **Depositional Landforms** folder.

2. Double-click each placemark in the **Depositional Landforms** folder to view examples of coastal landforms related to depositional processes.

Exploration 20.2 Coastal Landforms Part B

B. Which of the placemarked locations in the **Depositional Landforms** folder illustrates a lagoon?

    1. A
    2. B
    3. C
    4. D
    5. E

Exploration 20.2 Coastal Landforms Part C

C. Which of the placemarked locations in the **Depositional Landforms** folder illustrates a tombolo?

    1. A
    2. B
    3. C
    4. D
    5. E

When you complete these parts, turn off and collapse the **Depositional Landforms** folder.

Instructions for Exploration 20.2 Coastal Landforms Part D:

1. Turn on and expand the **Beaches** folder.

2. Double-click each placemark in the **Beaches** folder to view examples of beaches with varying composition. Be sure to note the relative slope of the placemarked beach locations.

Exploration 20.2 Coastal Landforms Part D

D. Which beach slopes of the placemarked locations in the **Beaches** folder is the most likely to have the lowest gradient and thus the smallest size beach particles?

    1. A
    2. B
    3. C
    4. D
    5. E

When you have completed this exploration, turn off and collapse the **20.2 Coastal Landforms** folder.

Exploration 20.2 – Coastal Landforms
Short Answer

Instructions for all Parts:

1. Make sure you have opened the **KMZ** file from www.mygeoscienceplace.com.

2. From the Places panel, expand **20. Coastal Processes and Terrain**.**kmz** and then open the **20.2 Coastal Landforms** folder.

Instructions for Exploration 20.2 Coastal Landforms Short Answer A:

1. Turn on and double-click the **Chesapeake Bay** folder. This imagery illustrates the bathymetry, or underwater topography, of the Chesapeake Bay estuary.

2. Study the contours of the underwater topography.

Exploration 20.2 Coastal Landforms Short Answer A

A. Describe the bathymetry of Chesapeake Bay. Are old river channels visible? If so, are the relatively deeper channels continuous or broken? Explain the processes that could create this underwater landscape.

_____
_____
_____
_____
_____

When you complete this part, turn off the **Chesapeake Bay** folder.

Instructions for Exploration 20.2 Coastal Landforms Short Answer B:

1. Turn on and expand the **Depositional Landforms** folder.

2. Double-click each placemark in the **Depositional Landforms** folder to view a collection of coastal landforms related to depositional processes.

Exploration 20.2 Coastal Landforms Short Answer B

B. It is rare for a coastal depositional feature to occur as a singular depositional landform upon the landscape. Select one of the landscapes associated with the placemarked landforms in the **Depositional Landforms** folder and briefly list and describe all of the depositional landforms present at that location.

_____

_____

_____

_____

_____

When you have completed this exploration, turn off and collapse the **20.2 Coastal Landforms** folder.

Exploration 20.3 – Coral Coasts
Multiple Choice

In the world's tropical oceans, many landmasses are fringed by coral structures. These living landforms provide important habitats for sea-dwelling creatures and a measure of protection for tropical coastlines. Many of the world's coral ecosystems are experiencing degradation due to coral bleaching that occurs with higher sea temperatures.

Instructions for all Parts:

1. Make sure you have opened the **KMZ** file from www.mygeoscienceplace.com.

2. From the Places panel, expand **20. Coastal Processes and Terrain.kmz** and then open the **20.3 Coral Coasts** folder.

Instructions for Exploration 20.3 Coral Coasts Parts A–C:

1. Turn on and open the **Coral Landforms** folder.

2. Double-click each placemark to view distinct landforms found along coral coasts.

Exploration 20.3 Coral Coasts Part A

A. Which of the placemarked features in the **Coral Landforms** folder illustrates a fringing reef?

1. A
2. B
3. C
4. D
5. E

Exploration 20.3 Coral Coasts Part B

B. Which of the placemarked features in the **Coral Landforms** folder illustrates an atoll?

1. A
2. B
3. C
4. D
5. E

Exploration 20.3 Coral Coasts Part C

C. Which of the placemarked features in the **Coral Landforms** folder illustrates a barrier reef?

1. A
2. B
3. C
4. D
5. E

When you have completed these parts, turn off and collapse the **Coral Landforms** folder.

Instructions for Exploration 20.3 Coral Coasts Part D:

1. Turn on and open the **Global Coral** folder.

2. Double-click each placemark to view five marine locations, noting the latitude and longitude of each site.

Exploration 20.3 Coral Coasts Part D

D. Based on location alone, which of the placemarks in the **Global Coral** folder is **not** a coralline structure?

1. A
2. B
3. C
4. D
5. E

When you have completed this exploration, turn off and collapse the **20.3 Coral Coasts** folder.

Exploration 20.3 – Coral Coasts
Short Answer

Instructions for all Parts:

1. Make sure you have opened the **KMZ** file from www.mygeoscienceplace.com.

2. From the Places panel, expand **20. Coastal Processes and Terrain.kmz** and then open the **20.3 Coral Coasts** folder.

Instructions for Exploration 20.3 Coral Coasts Short Answer A:

1. Double-click the *Reef Scene* Gigapan image to view an underwater image of a coral reef.

2. Study the image, then use the navigation tools in Google Earth™ to evaluate the location where the image was captured.

Exploration 20.3 Coral Coasts Short Answer A

A. Based on the *Reef Scene* image and the location where the image was captured, describe the potential economic value of this site.

_____

_____

_____

_____

_____

Instructions for Exploration 20.3 Coral Coasts Short Answer B:

1. Turn on and open the **Coral Landforms** folder.

2. Double click the *A* placemark to view Bikini Atoll.

3. Utilize outside resources to study the cultural significance of this landform.

Exploration 20.3 Coral Coasts Short Answer B

B. What is the cultural significance of Bikini Atoll? What locational and physical characteristics contributed to the selection of this site for its historically significant activities?

_____

_____

_____

_____

_____

When you have completed this exploration, turn off and collapse the **20.3 Coral Coasts** folder.

Exploration 20.4 – Human-Coastal Dynamics
Multiple Choice

The coastal interface is increasingly affected by humans and their activities. In coastal areas, humans introduce pollutants, actively manage sediments, and modify natural landforms. The ubiquity of humans in coastal spaces will continue to influence the natural processes within this zone.

Instructions for all Parts:

1. Make sure you have opened the **KMZ** file from www.mygeoscienceplace.com.

2. From the Places panel, expand **20. Coastal Processes and Terrain.kmz** and then open the **20.4 Human-Coastal Dynamics** folder.

Instructions for Exploration 20.4 Human-Coastal Dynamics Part A:

1. Turn on and double-click the *Japan Tsunami* placemark.

2. Evaluate the human and physical landscape based on the 9/20/2010 imagery.

3. Use the time slider to view the imagery from 3/11/2012. View the subsequent images until you reach the most recent imagery.

Exploration 20.4 Human-Coastal Dynamics Part A

A. Which of the following statements is most strongly supported by the historical imagery associated with the *Japan Tsunami* placemark?

1. By 2/02/12, the area around the *Japan Tsunami* placemark had been completely rebuilt.
2. All of the standing water in the fields had dried and/or evaporated by 3/16/2011.
3. The tsunami did not have an impact on the beach sediments located east of the *Japan Tsunami* placemark.
4. No local vegetation survived the tsunami.
5. At this location, evidence of tsunami flooding is evident more than three kilometers inland.

IMPORTANT: Be sure to turn off the Historical Imagery.

When you have completed this part, turn off the *Japan Tsunami* placemark.

Instructions for Exploration 20.4 Human-Coastal Dynamics Part B:

1. Turn on and open the **Jetty** folder

2. Double-click placemarks *A* through *E* to view examples of coastal landscapes.

3. Identify the placemarked feature that is the best example of a jetty.

Exploration 20.4 Human-Coastal Dynamics Part B

B. Which placemarked feature in the **Jetty** folder is the best example of a jetty?

1. A
2. B
3. C
4. D
5. E

When you have completed this part, turn off and collapse the **Jetty** folder.

Instructions for Exploration 20.4 Human-Coastal Dynamics Part C:

1. Turn on and double-click the *Managing Coastal Sediment* placemark.

2. Evaluate the placemarked feature.

Exploration 20.4 Human-Coastal Dynamics Part C

C. What is the anthropogenic feature marked by the *Managing Coastal Sediment* placemark?

1. Jetty
2. Groin
3. Longshore drift
4. Tombolo
5. Spit

When you have completed this part, turn off the *Managing Coastal Sediment* placemark.

Instructions for Exploration 20.4 Human-Coastal Dynamics Part D:

1. Turn on and double-click the **Human Impacts** folder.

2. Evaluate the cumulative impact of stressors associated with human impacts on marine ecosystems in coastal areas. Heavily impacted areas are colored orange and red while relatively less impacted areas are shown as light green and blue.

Exploration 20.4 Human-Coastal Dynamics Part D

D. Which of the following statements is most strongly supported based on an assessment of the data in the **Human Impacts** folder?

1. The greatest human impacts are concentrated near the equator.
2. Coastal areas of East Asia do not exhibit significant human impacts according to this dataset.
3. California coastal areas exhibit the highest levels of human impacts for the contiguous United States.
4. Coastal areas of the South Atlantic Ocean are relatively more impacted than coastal areas of the North Atlantic Ocean.
5. Coastal areas in the northern hemisphere have relatively higher levels of human impacts than coastal areas in the southern hemisphere.

When you have completed this exploration, turn off and collapse the **20.4 Human-Coastal Dynamics** folder.

Exploration 20.4 – Human-Coastal Dynamics
Short Answer

Instructions for all Parts:

1. Make sure you have opened the **KMZ** file from www.mygeoscienceplace.com.

2. From the Places panel, expand **20. Coastal Processes and Terrain.kmz** and then open the **20.4 Human-Coastal Dynamics** folder.

Instructions for Exploration 20.4 Human-Coastal Dynamics Short Answer A:

1. Turn on and double-click the *Long Beach* placemark to view an area of significant coastal alteration by humans.

2. Study the image, noting likely anthropogenic alterations.

Exploration 20.4 Human-Coastal Dynamics Short Answer A

A. Describe the alterations to the coastline that are visible near the *Long Beach* placemark. What are the anthropogenic features that have been created? Are "natural" coastal landforms evident?

_____

_____

_____

_____

_____

Instructions for Exploration 20.4 Human-Coastal Dynamics Short Answer B:

1. Turn on and double-click the *Port Mansfield* placemark to view an area of coastal alteration by humans.

2. Study the image, noting likely anthropogenic alterations.

Exploration 20.4 Human-Coastal Dynamics Short Answer B

B. What linear feature(s) are shown on either side of the *Port Mansfield* placemark? Why have these features been constructed? What is their relation to the broader landscape seen by zooming out?

_____

_____

_____

_____

_____

When you have completed this exploration, turn off and collapse the **20.4 Human-Coastal Dynamics** folder.